Holistic Product Development

Challenges in Interoperable Processes, Methods and Tools

edited by
Prof. Dr. Werner Dankwort
University of Kaiserslautern

based on
the International 5th Workshop
on Current CAx-Problems
held at Maria Rosenberg (Kaiserslautern),
Germany on April 5th – 7th, 2004

Berichte aus der Informationstechnik

Werner Dankwort (Ed.)

Holistic Product Development

Challenges in Interoperable Processes,
Methods and Tools

Shaker Verlag
Aachen 2005

Bibliographic information published by Die Deutsche Bibliothek
Die Deutsche Bibliothek lists this publication in the Deutsche
Nationalbibliografie; detailed bibliographic data is available in
the internet at http://dnb.ddb.de.

Copyright Shaker Verlag 2005
All rights reserved. No part of this publication may be reproduced, stored in a
retrieval system, or transmitted, in any form or by any means, electronic,
mechanical, photocopying, recording or otherwise, without the prior permission
of the publishers.

Printed in Germany.

ISBN 3-8322-3855-7
ISSN 1610-9406

Shaker Verlag GmbH • P.O. BOX 101818 • D-52018 Aachen
Phone: 0049/2407/9596-0 • Telefax: 0049/2407/9596-9
Internet: www.shaker.de • eMail: info@shaker.de

Preface

This workshop was hold in remembrance of our colleague and friend Josef Hoschek.

Prof. Josef Hoschek first founded this workshop in 1993. With fine tradition this years' workshop was the fifth and it was held in honour of Prof. Hoschek since he passed away in 2002. As the founder of these workshops he swayed the first three workshops by his ideas and his personality. Unfortunately he could not participate in the fourth workshop, but his fundamental ideas can still be seen in the workshop's intention.

It was always his thinking to combine science, especially mathematics, with the field of application. To mention only two activities, in which some of us from the current CAx-Workshop were engaged: Josef Hoschek was one of the founders of the famous Oberwolfach CAGD-Meetings, which was regularly hold since the beginning of the 80'tie over many years and he was also involved in various projects with the Association of the German Automotive Industry (VDA).

The intention of the CAx-workshops was to bring the different groups of IT supported engineering together: industrial product development and manufacturing, system suppliers and research. As we all know, the information flow between competing enterprises and companies is often blocked, even though objectively in many areas the common interest of shared knowledge should dominate and would bring advantages to all of us. At these workshops we tried to generate an atmosphere of mutuality, to intensify personal relations, so that common objectives will be discussed and even solutions for common problems or challenges will be achieved.

In today's industry the product development moved more and more to virtual reality – the new slogan is "Virtual product". The classical gap between IT-thinking and product development is not yet closed. Due to today's situation to bridge that gap should be one of the most important goals today and in the future. Since the beginning of CAD many years ago an important problem was (and currently is) the data exchange. Now the challenge of the daily increasing information flow, not only data exchange, has an extreme impact on the efficiencies of the industrial workflow.

Therefore this question of interoperability has to be treated on various levels:
Many different IT systems with quite different abilities are necessary to support the

industrial processes. The interoperability between these systems, their data, and related methods is the huge technical task. – Interoperability within the industrial product definition and generation processes between the various steps and sub-processes leads to interoperability between the companies, the organisations and even between the persons. Monopolistic IT-structures sometimes promise to bring the solution of all interoperability problems, but any one in industrial life knows about the reality. The pressure of the global market requires innovations in all fields: in products – in processes – in methods – in IT-support – in IT-tools: However there is not only a large field of research regarding the systems and methods, we are also challenged by personal cooperation, enterprises, personal level and by human culture.

The global objectives of this workshop was not only to find approaches for technical solutions of actual issues, but also to get some ideas of future trends, how we can be prepared for what the future will bring, or how we may have impact on long term tendencies.

In this workshop there were colleagues from various areas of the industry – automotive and suppliers, from companies supporting industrial processes, from software vendors, and – last not least – from research and education. They came from various countries: from Europe, Israel, USA and Japan. The topics of this workshop were: Industrial processes and workflow, product definition, IT-aspects, methods, tool, fundamentals, PLM, as well as strategic and legal aspects.

We hope that the added value, every one gained from this workshop, will go beyond acquired knowledge of the technical information, and that the personal relations to the colleagues of other working fields, other companies or institutes will be intensified.

Kaiserslautern, 2004   C. Werner Dankwort

Josef Hoschek   1935 - 2002

Contents

Preface	3
Contents	7
Part I: Keynote Paper	10
Nowacki, Horst:	
On Methodologies for Product Optimization	11
Part II: Industrial Processes & Workflow	30
Katzenbach, Alfred; Feltes, Michael; Willis, Joanne:	
Collaborative Engineering – A Challenge for More Effective Engineering	31
Klamann, Rolf:	
Engineering Change Management as a Part of Electronic RFQ	47
Kraus, Peter:	
Generative Car & e-Vehicle	51
Shimomura, Daisuke:	
Nissan Digital Process and Next Steps	63
Stark, Rainer:	
PLM in Automotive Engineering - Today's Capabilities & Tomorrow's Challenges -	71
Takagi, Atsushi; Hiroaki, Shimada:	
Digital Colour Process for Automobile Design	81
Part III: Product Definition	104
Ohtaka, Akihiko:	
Status of the ISO TC184/SC4 Parametrics Standards Development	105
Ovtcharova, Jivka:	
Living Vehicles – The Paradigm Shift towards Holistic Product Creation in the Automotive Industry	121
Sapidis, Nickolas S.:	
Geometric Modeling of Layout Constraints for Plant PLM (or New Challenges for Solid Modeling imposed by PLM)	133
Schneider, Franz-Josef:	
The "Digital Factory"	151
Suzuki, Hiromasa:	
Convergence Engineering	157
Weber, Christian:	
Modern Products: New Requirements on Engineering Design Elements, Development Processes, Supporting Tools and Designers	165

Part IV: IT Aspects, Methods & Tools   180
Allen, George:
    The Evolution of New PLM Technologies   181
Blair, David:
    Product Development System with Special Focus on Automotive   195
Cremers, Anton, M.; Reindl, Peter
    Software Systems Development and Deployment   203
Massabo, Alain:
    Any Future for Parametric Design?   213
Trautheim, Andreas; Millarg, Ludger:
    Long Term Archiving and Retrieval (LOTAR) of Digital Product Data   221

Part V: Fundamentals   238
Andersson, Roger K. E.:
    Surface Design through Modification of Shadow Lines   239
Andersson, Roger K.E.; Johansson, Bo I.:
    Representation of Intersection Curves   257
Bercovier, Michel; Luzon, Moshe; Pavlov, Elan:
    Reverse Engineering of Objects Made of Algebraic Surfaces Patches   277

Part VI: PLM   292
Cugini Umberto; Bordegoni Monica:
    From Systematic Innovation to PLM   293
Laufer, Philippe:
    3D PLM Innovation through Openness   303

Part VII: Strategic and Legal Aspects   316
Salzmann, Peter:
    Embedding Niche Software into Heterogeneous CAD Environments   317
Stopper, Martin:
    Interfaces as Essential Facility and EC Antitrust Law   331
Dankwort, Werner:
    Outlook for the CAx/PLM Future   339

Panel Discussion (Summary)   351

Part I: *Keynote Paper*

On Methodologies for Product Optimization

Horst Nowacki

Technische Universität Berlin, Germany

Abstract: A major goal of product development by CAx systems has been the improvement of product quality by evaluation of rational criteria and by automatic or semi-automatic optimization methods. The level of achievement in supporting these objectives has steadily improved over several decades, the criteria and goals for holistic design have become more and more ambitious. Yet a definitive approach has not yet evolved and many hopes were disappointed. This presentation will give a broad assessment of key developments in problem formulations based on optimization methodologies and will seek to evaluate achievements and deficits. Examples will be drawn from product development in shipbuilding, especially in hull form design. The usefulness of a holistic design approach, based on a product lifecycle perspective and embedded in a teamwork scenario of concurrent engineering will be demonstrated.

1 Introduction

This opening paper, presented at the 5th International Workshop on Current CAx-Problems, is setting the stage for the main theme of the meeting: "Holistic Product Development – Challenges in Interoperable Processes, Methods and Tools". It will serve to offer a definition of what we mean when we say that the product development shall be *holistic*. The term "holistic" is perhaps fashionable, but also a bit vague unless placed in its proper context.

Literally "holistic" (from Greek 'όλον = whole) just means pertaining to the whole, all-embracing. In the context of product development a *holistic* design process refers to the _whole product in *all* of its aspects *during the entire lifecycle*.

In my understanding "holistic development" thus requires that it be based on a *unified perspective of goals and restrictions for the entire decision process.* This unified perspective can then be shared by all participants in the development process in the modern scenarios of design teamwork with multiple concurrent decision-makers. It offers a communication platform, based on rational, shared criteria, for harmonizing decisions made in design organisations with an industrial division of labor.

In my presentation I consider optimization methodologies as tools in holistic design. Optimization problems are stated formally by making the quality criteria (measures of merit) and restrictions (constraints) explicit. They are solved by searching for the best solution to this set of requirements.

Thus the optimization format is one specific way of looking at holistic design. Hence indirectly by evaluating the achievements of optimization in design we can assess also the success stories and deficits in holistic design decision-making (Nowacki, [1]).

This is what I will try to address here. Although my discussion of problem formulations will be application independent, my illustrative examples will be drawn from my field of ship design.

2 Optimization Problem Statement

The object of optimization in product development is usually a product and/or process, designated as a *system*. The system is to be optimized on the basis of a rational criterion, called the *measure of merit* and in face of the given restrictions, called the *constraints*. Let us denote (Fig. 1):

D = vector of free variables (design variables)
P = vector of parametric influences
M (**D**, **P**) = measure of merit function
C (**D**, **P**) = vector of given constraint functions

Then, in accordance with the input/output information flow in Fig. 1, the optimization problem can be stated in the standard format:

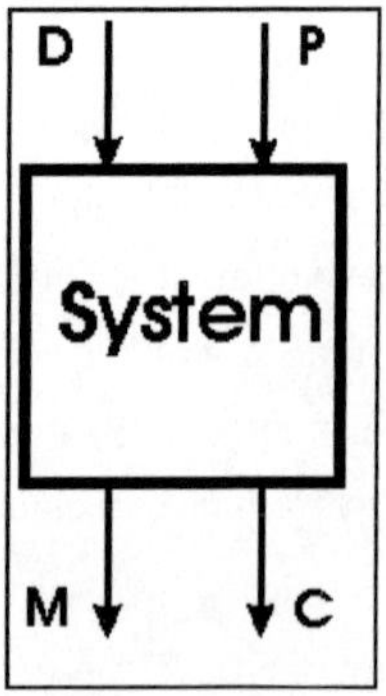

Figure 1.: System to be Optimized

" Find the set of design variables $\mathbf{D}^*$ for given parameters $\mathbf{P}^*$

that will minimize (maximize) the measure of merit:

$$M(\mathbf{D}^*, \mathbf{P}^*)$$

and satisfy the set of J given constraints:

$$C_j(\mathbf{D}^*, \mathbf{P}^*) \geq 0 \quad \text{for } j = 1, ..., J. \text{ "}$$

This problem type is generally known as *Nonlinear Programming*, provided that either the measure of merit function or at least one of the constraint functions are nonlinear expressions in the design variables, which is the predominant case in design applications. The constraints are here regarded as inequalities without loss of generality because any existing equality constraints are indirectly taken into account either by allowing for them in the calculation of the measure of merit and constraints or by replacing each equality constraint by two inequalities.

In modelling a design task as an optimization problem it is important to comply with certain rules:

Measure of merit:

The measure of merit is generally a *ratio* of *benefits* to related *efforts*. Benefit or effort *alone* is a legitimate choice *only if* the other influence is *constant* for all solutions.

- Preferably benefits and efforts should be modeled in the same units. The most convincing choice is some economic criterion where benefits and efforts are described in *monetary units*. This often requires conversion of technical performance measures into economic measures in a specific *economic model* where economic consequences are quantified. Holistic design which deals with several technical quality measures can hardly be achieved without such economic yardsticks. Some engineers shy away from economic comparisons, but the difficulties of quantification are usually not prohibitive once the skills of economic modeling are acquired. The economic model for a technical decision helps to establish commensurate units in complex decision processes.
- Formulations with a *single measure of merit* in practice are much to be preferred to problem statements with *multiple merit criteria* (multi-objective optimization). It is not rare that a design must meet several functional objectives, but this does not per se necessitate multiple measures of merit. Rather some of the objectives are simply constraints, the others can frequently be combined into a single criterion if they can be expressed in

economic or other commensurate units. Only if this should really fail, one can still resort to multi-criteria optimization (Pareto optimization).

The decisions that lead to a choice of measure of merit will necessarily reflect in whose interest the system is to be evaluated. This cannot be avoided, but must be clearly stated. E.g. different orientations on benefits and preferences exist between the system producer and the system operator. This difference cannot be resolved on objective grounds, but must be overcome by a subjective decision on the choice of the system beneficiary. Thus even a unified, "holistic" design perspective does *not* ensure a *unique* statement of the optimization objectives. This should be understood and made transparent.

The recommendation of favouring economic comparisons of product development alternatives must not be misunderstood as an advocacy of "return on investment" as the sole, only saving faith. On the contrary the concept of holistic product development, if suitably implemented, creates the perspective of pursuing a concerted process of simultaneous product improvements in many categories. Future products and services thus shall holistically become more:

- Functionally efficient
- Safe, fail-proof, long-lasting
- Multifunctional, flexible
- Affordable
- Material and energy saving
- Space and time saving
- Efficiently producible
- Clean, non-pollutant, recyclable
- Maintainable, repairable, disposable
- Stylish, beautiful, culturally valuable
- Creative, enriching
- Easy-to-use etc.

It is evident that this list contains many non-economic objectives, which however does not contradict the solutions being compared on economic grounds.

Constraints:

The rules for modeling constraints include:
- Enumerate *all* restrictions (functional, contractual, regulatory, social, human).

- For each constraint set target values:

$$C_{ACTUAL} <= C_{PERMISSIBLE}$$

- Evaluate the state the constraints by calculation, simulation or virtual reality mock ups.
- Monitor which constraints are active and assess the consequences of their violation.

3 Analysis of Problem Structure

As soon as the decision task is formulated as an optimization problem, the structure of the problem type can be analyzed. One of the major difficulties in holistic design is dealing with complex, large-size problems for which a direct solution as a single step optimization problem is often prohibitive in magnitude. But a single-stage solution is not inevitably required even if the problem statement has resulted from a unified perspective. Rather one should aim at a skilful *decomposition* into smaller, better manageable subtasks. This is not unlike the division of labor in a traditional design team, though guided by unified rational criteria.

The decomposition may take the following steps:

- Characterize problem type (linear/nonlinear; continuous/discrete design variables; single/multiple criteria etc.).
- Identify problem size (number of design variables, constraints, measures of merit).
- Delineate decision process structure (serial, parallel, stage wise etc.). A *stage wise system* is one whose decision process can be broken down into subsets, called *stages*, where each stage i has its own subset of design variables D_i and its own contribution M_i to the measure of merit (Fig. 2).
- Decompose decision process into stages, where possible.
- Sub optimize each stage as a function of its input state, then seek the optimal combination of stages, based on the approach of Dynamic Programming (Nemhauser [2]).
- Solve the problem numerically and update as the development process advances.
- Discuss existence and uniqueness of solutions (multimodality issues).

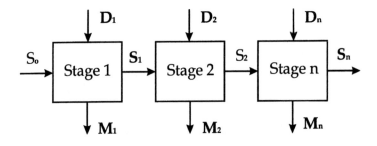

Figure 2: stage wise system structure

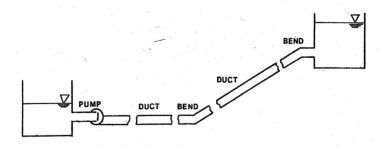

Figure 3: Reservoir-duct-bend piping system as stage wise technical system

A simple example of stage decomposition of a decision process is shown in Fig. 3 for a serial system of n stages, a duct-bend piping system, piping fluid from one reservoir level to another. This system physically consists of a pipeline through which fluid is moved by a pump. The pipeline has several elements (i) of straight ducts and bends whose dimensions (diameters D_i) are to be optimized. Each stage has its own measure of merit M_i, viz., its contribution to the investment and operating cost of the system over the lifecycle. As the pipe diameter increases, the former cost rises, the latter decreases and vice versa. The sum of all stage measures of merit $M = \Sigma\ M_i$ is to be minimized for the system. In Dynamic Programming each stage is first separately pre-optimized as a function of its input state S_i, which results from the upstream stages. These sub problems are small and all non-optimal operating conditions of the stages are thus eliminated in this step. Then the best combination of optimal stages is found by minimizing the measure M for the whole system. This "outer loop" of optimization is much smaller in problem size than for the whole system initially because only the pre-optimized, "optimally tuned" stages have remained in contention.

An analogous approach can be taken if the application does not deal with a physical system of several components, but rather with the logical process of stage wise decisions in product development. This way of structuring the development process into several stages is almost mandatory if the objective of *holistic design* for all development stages is taken seriously.

4 On Methods and Tools

In order to adopt a holistic approach to product development the modeling methods and software tools are largely available. The challenge lies in organizing an environment based on a unified understanding of objectives and supporting the necessary cooperation and communication within the development team.

The *methodology* will be based on:
- Monitoring of product development through all stages at the component/ subassembly/ assembly level, in particular monitoring of time and cost in each relevant sub process (by PLM software).
- Optimizing sub processes (by optimization software), then the total system.
- Focussing on time-dependent processes and optimizing event chains (like in concurrent engineering, material ordering, material flow control, scheduling, production etc.)

The *tools* will include software for:
- Product and process description (CAx, PLM)
- Assessment of weight, time, cost
- Physical simulation (digital mock ups, virtual reality)
- Operational simulation (operations research, queuing analysis)
- Automation in production (robotics, simulation and control)
- Concurrent engineering (synchronization, coordination)
- Optimization

5 Examples

5.1 A "Holistic" Design Criterion for Ships

The ship owner from an economic viewpoint will judge the performance of a cargo ship by its transport capacity per unit of invested and operational cost. This is not unlike the approach taken by other transport enterprises, e.g., the passenger airlines when they evaluate lifecycle cost per passenger-mile or similar measures.

More precisely for a cargo ship the owner can go by a holistic criterion, called *Ship Merit Factor* (SMF), introduced by Cheng [3], and defined by:

Ship Merit Factor SMF (in net ton-miles/ €) =

= net annual transport capacity (ton-miles/yr.)/ average annual cost (€/yr.),

hence

$$SMF = (W_{PL} \ V_S) \ f_S \ f_L \ f_V \ f_{ST} / (AAC),$$

where

W_{PL} = design payload (tons)

V = design speed (knots)

AAC = average annual capital cost plus operating cost (€)

f_S = service time factor (uptime percentage)

f_L = load factor (net payload/design payload)

f_V = operating speed/ design speed

f_{ST} = sea time/ year (percent),

and with

η_P = propulsive efficiency

R_T = total resistance of ship

Δ = displacement (tons)

P_B = installed engine power (HP)

const = a unit conversion factor

this can be expanded into:

$$SMF = \text{const} \ (W_{PL}/ \Delta)(\Delta / R_T) \ P_B \ \eta_P \ f_S \ f_L \ f_V \ f_{ST} / (AAC)$$

The SMF criterion is holistic by virtue of its overall economic evaluation of the ship in its lifecycle for its principal cargo transport function, but it is at the same time decomposed into technical and operational performance measures which can serve as stage wise criteria or departmental design objectives and operational goals. Thus each participant in the decision process has a clear understanding of how he may contribute to the ship owner's overall success.

5.2 Form Parameter Design of Ship Hull Surfaces

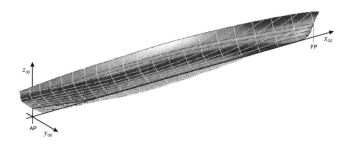

Figure 4: Motor yacht Hull Shape to be Designed (from [4])

The design of a ship hull surface from specified form parameters is a task of geometric modeling with equality constraints on certain properties of the hull geometry. This task can be approached as an optimization problem where a *fairness functional* is introduced to serve as a measure of merit. For example for a motor yacht hull shape to be designed (Fig. 4), following an approach outlined by Stefan Harries [4], [5], the designer poses the following the problem:

"Find a hull surface minimizing a fairness functional while complying with given form data on volumes, volume centroids and constraining curves of the surface!"

Thus the process starts with form parameters, works out the constraining *"basic curves"* and then solves for a surface interpolating the basic curves while minimizing a surface fairness functional. The last step in this process is called *"fair skinning"*. See Figs. 5 and 6 for the overall process and its three stages.

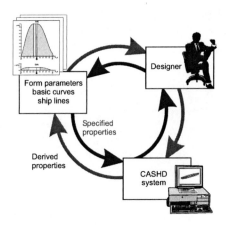

Figure 5: Design Cycle in Form Parameter Design, Following the Inner Loop (from [4])

Note that in ship design the underwater hull surface is subject to volume and centroid constraints for functional rea-sons, i.e., this surface domain has its enclosed volume and volume centroid prescribed. This is achieved by designing a planar curve, the Sectional Area Curve (SAC), Fig. 7, whose ordinates are the cross sectional areas of the underwater volume so that the integral of this curve defines the volume spanned by the underwater surface as well as its centroid. The SAC is a basic curve derived from given form parameters. Other basic curves are the ship's design waterline and its lateral profile, i.e., its centerplane contour. For this set of basic curves transverse section curves through the hull surface can be derived (Fig. 8). These cross sections inherit a prescribed area from the SAC, a draft and a breadth from other basic curves, but are still free in other parameters, e.g., the section flare angle at the design waterline level.

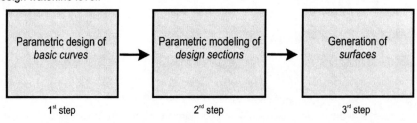

Figure 6: Stage wise Process of Hull Shape Form Parameter Design (from [4]

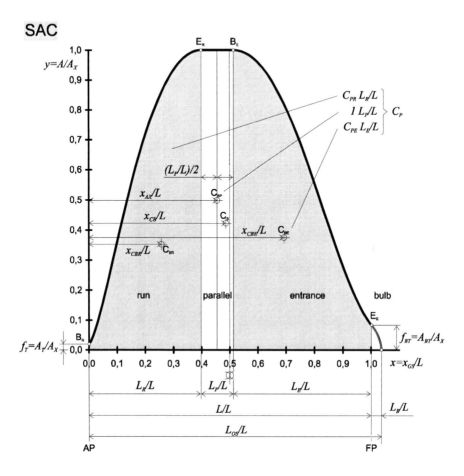

Figure 7: Generic Sectional Area Curve: Underwater Cross Sectional Areas A, Nondimensionalized by Midship Section Area A_X, against Length x, Nondim. by Ship Length L. Quantities denoted by symbols are specified form parameters (from [5]).

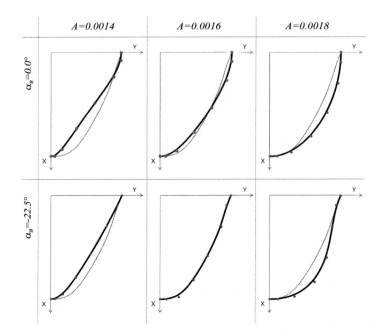

Figure 8: Systematic Variation of Forebody Section Shapes with Section Area A and Flare Angle α (from [4]).

The *"fair skinning"* process (Harries [4]) assumes a B-spline representation of the surface

$$\vec{F}(u,v) = \begin{pmatrix} x(u,v) \\ y(u,v) \\ z(u,v) \end{pmatrix} = \sum_{i=0}^{m-1}\sum_{j=0}^{n-1} \vec{P}_{ij} \cdot N_{ik}(u) \cdot N_{jl}(v)$$

and interpolates the given transverse sections as "skin curves" so that for each parameter v_d (at station x_d) the surface equals the given skin curve:

$$\vec{F}(u,v_d) = \sum_{i=0}^{m-1}\left(\sum_{j=0}^{n-1} \vec{P}_{ij} \cdot N_{jl}(v_d)\right) \cdot N_{ik}(u) \stackrel{!}{=}$$

$$\stackrel{!}{=} \sum_{i=0}^{m-1} \vec{V}_{id} \cdot N_{ik}(u) = \vec{Q}_d(u)$$

Consider the inner summation as a function of v

$$\vec{L}_i(v) = \sum_{j=0}^{n-1} \vec{P}_{ij} \cdot N_{jl}(v)$$

as a B-spline curve and allow enough free vertices P_{ij} (greater number than skin curve constraints). Then each curve $L_i(v)$ can be faired (optimized) by minimizing the second order fairness measure

$$E_{2_i} = \int_{v_B}^{v_E} \left\{ \left(\frac{d^2 x_i}{dv^2}\right)^2 + \left(\frac{d^2 y_i}{dv^2}\right)^2 + \left(\frac{d^2 z_i}{dv^2}\right)^2 \right\} dv$$

while, meeting the interpolation constraints at each skin curve (v_d). Thus the surface will contain all specified design sections and still be as fair as the constraints will permit. The process is automatic except for the required form parameter inputs and possibly a few adjustments of default settings. If the generated shape does not meet all explicit and implicit intentions, the form parameters must be changed. It is of course possible to change only one form parameter at a time, thus creating a univariate family of shapes.

A second example, based on the thesis of H.C. Kim [6], deals with the generation of a tanker hull form consisting of five surface and volume sub domains: The mid body, the bow and stern bulbs and two G1 continuous blending surfaces in the transition from bulbs to mid body. Fig. 9 shows the separate domains, Fig. 10 illustrates the surface decomposition. The local sub domains each have their own sets of form parameters and, when they are combined, the global hull form parameters are acting as constraints. Thus the same principle of form parameter design can be applied also to complex shapes with several domains. Fig. 11 presents the final hull form of a tanker (H.C. Kim [6]).

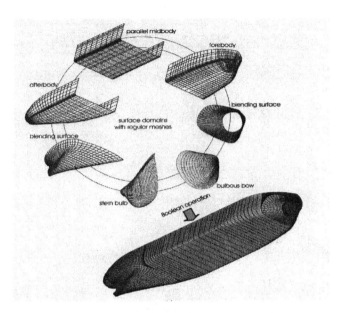

Figure 9: Decomposition of Tanker Hull Surface and Volume into Several Domains (from H.C. Kim [6])

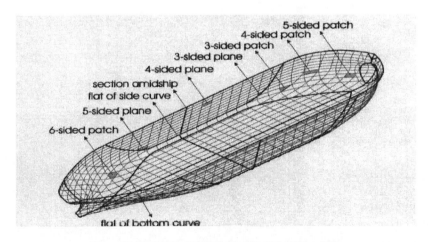

Figure 10: Tanker Hull Multiple Domain Subdivision ([6])

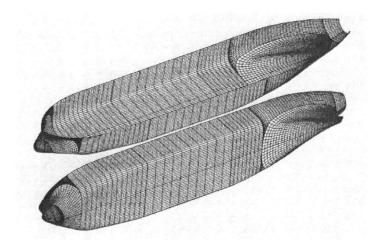

Figure 11: Form Parameter Designed Tanker Hull Form, Final Results ([6])

5.3 Hydrodynamic Hull Shape Design

The two examples in the preceding section are based strictly on geometric criteria, viz., form parameters, fairness measures and geometric constraints. For the ship owner the hydrodynamic performance of the vessel is probably the most important quality criterion for a hull form. Thus in a holistic development context the measure of merit for shape evaluation should be of hydrodynamic orientation. This is achieved by using a hydrodynamic criterion, e.g., resistance per ton of displacement (R_T/Δ), as measure of merit at this stage of design, as discussed in Section 5.1.

The design process, as demonstrated by Harries [4], is then organized as a cycle of four stages (Fig. 12):

- o Parametric form generation
- o Hydrodynamic performance analysis of a given shape by means of Computational Fluid Dynamics systems
- o Evaluation of the hydrodynamic measure of merit
- o Form parameter variation by an optimization strategy

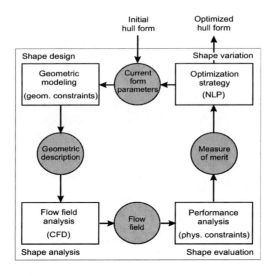

Figure 12: Optimization Cycle in Hydrodynamic Design ([4])

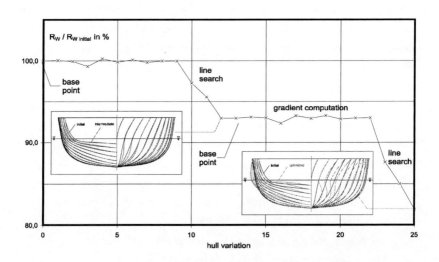

Figure 13: Results of Hydrodynamic Shape Optimization Run ([4]).

Thus the geometric form generation capability presented in Section 5.2 is now embedded in a greater loop in which the hydrodynamic performance is evaluated and optimized by shape variation within the given geometric constraints. Fig.13 shows the results of an optimization run for a high-speed round-bottom hull form.

5.4 Summary

The set of examples presented in the preceding subsections will serve to illustrate an approach in which optimization sub problems are formulated to be subservient to an overall holistic design approach. The holistic criterion was broken down into stage wise criteria for a set of sub problems (Section 5.1). The subtask of form generation by form parameters was treated by an optimization process minimizing a fairness functional, though not from a holistic viewpoint (Section 5.2). But the form generation module was then placed in the context of a hydrodynamic optimization process (Section 5.3) whose measure of merit is hydrodynamically oriented and thus a stage wise sub factor in the holistic general balance. Similar methods of problem decomposition can be applied to the other stages of the design process. The stage contributions can be monitored and should be reassembled in a holistic overall design evaluation.

6 Conclusions

The "holistic" design approach of the future requires rational, quantifiable product quality criteria in order to make systematic improvements over the development and operational lifecycle. Future systems must be capable of assessing technical performance, time and cost requirements at every relevant stage of the development cycle. Optimization can play an increasing role in structuring development processes and in systematically exploiting improvement potentials. This was demonstrated also by some examples from ship design in this paper. Product development in many industries is embedded in concurrent engineering, highly communicative environments within and between companies. Holistic design in such scenarios is feasible only by relying on the synergies in interoperable processes, methods and tools.

6.1 Acknowledgment

The author is pleased to acknowledge his gratitude for the permission to quote several examples from the excellent theses by Stefan Harries and Hyun Cheol Kim.

6.2 References

[1] Horst Nowacki, "Design Synthesis and Optimization – An Historical Perspective", invited paper, 39th WEGEMT Summer School OPTIMISTIC – Optimization in Marine Design, Proc. edited by Lothar Birk and Stefan Harries, Mensch & Buch Verlag, Berlin, 2003.
[2] G. Nemhauser, "Introduction to Dynamic Programming", John Wiley & Sons, New York, 1966.
[3] Henry M. Cheng, "Performance Comparisons of Marine Vehicles", SNAME New York Metropolitan Section Paper, Sep 1968.
[4] Stefan Harries, "Parametric Design and Optimization of Ship Hull Forms", Ph.D. Thesis, Techn. Univ. Berlin, Mensch & Buch Verlag, Berlin, 1998.
[5] Stefan Harries and Horst Nowacki, „Form Parameter Approach to the Design of Fair Hull Shapes", Proc. 10th International Conf. on Computer Applications in Shipbuilding (ICCAS), Proc. edited by C. Chryssostomidis and K. Johansson, MIT Sea Grant College Program, Cambridge, MA, 1999.
[6] Hyun Cheol Kim, "Parametric Design of Ship Hull Forms with a Complex Multiple Domain Topology", Ph.D. Thesis, Techn. Univ. Berlin, Mensch & Buch Verlag, Berlin, 2004

7 Author

Horst Nowacki, Prof. em. Dr.-Ing., Technische Universität Berlin

nowacki@ism.tu-berlin.de

Part II: *Industrial Processes & Workflow*

Collaborative Engineering – A Challenge for More Effective Engineering

Katzenbach, A. [1]; Feltes, M. [2]; Willis, J. [3]

1) Daimler Chrysler AG, Ulm, Germany
2) Daimler Chrysler AG, Ulm, Germany
3) Protics GmbH, Stuttgart, Germany

Abstract: Collaborative Engineering aims at increasing engineering efficiency in the automotive industry across different companies and countries. It aims to change the way people work so that new potential is created which can levered by automotive companies. This contribution outlines the need for Collaborative Engineering within the automotive industry as well as the associated risks. It also outlines the changes required to align the current situation and the general requirements for Collaborative Engineering.

1 The need for change

The automotive industry is a highly competitive industry and companies within it are continuously looking for new ways for economic growth. Current trends show that this is often done by increasing revenue, for example, through an expansion of existing markets, the entering of new markets, the production of niche products or through an increase in productivity. However, because of the complexity of these measures the potential for economic growth is only moderate.

As an alternative, the automotive industry could look at increasing efficiency as a way of economic growth. This could simply be done by reducing the variance of the different parts in vehicles. Why, for example, should several windscreen wipers be developed if one type can be made to fit several vehicles. Or taken on a larger scale, a company producing similar parts in different world-wide locations may be able to produce one part in one location for several vehicles. In addition, the use of platforms for different vehicle models and brands as well as the use of partner-overlapping networked competencies and resources all lead to an increase in efficiency for the automotive industry. As these methods are less complex and less expensive than the methods for increasing revenue, the potential for economic growth is extremely high.

This increase in efficiency leads to a competitive advantage for companies and one of the ways in which this can be achieved is through the use of Collaborative Engineering. Collaborative Engineering lets distributed engineering teams work together by providing common technologies and processes. It is especially useful in the automotive industry where development plants are in distributed locations, often across the globe.

Collaborative Engineering brings many benefits including increased productivity and efficiency as resources in several locations can be combined. For example, development activities can be analyzed individually in order to decide how resources can be best combined according to their level of importance and/or complexity. Innovation and creativity also increase due to the creation of new teams, sometimes with very different cultures or approaches [1].

However, Collaborative Engineering also has some risks, which should be taken into consideration by companies planning to go ahead with Collaborative Engineering. It provides people with a new way of working which, at the start, could lead to lower work satisfaction. In addition, Collaborative Engineering requires people who speak different languages to communicate with each other so communication can be slow and difficult to start with. People with excellent technical knowledge may not be able to use the common language so translators may be required. This again hinders communication and some people may feel that their message is not being correctly brought across, especially when technical details are being described. In the worst case, all of these factors can leave a person feeling unsatisfied and reluctant to participate. Nevertheless, Collaborative Engineering does have its benefits and, providing the risks mentioned above are overcome, can lead to businesses gaining a large competitive advantage. The risks themselves could be overcome by a clear and open communication strategy pointing out the benefits and long term results of such an initiative. The company should also point out medium term goals for the organization such as the development of common standards and processes as well as the development of new, easy to use and capable IT solutions so that all employees are able to see how a new initiative can be put into place.

With Collaborative Engineering several products can be created in parallel by various partners at different locations. The collaboration project team can be made up of a number of partners who all use common data, data structures and processes and who all have joint online access to product data. Briefly said, Collaborative Engineering is a liaison between partners. Collaborative Engineering should lead to improved quality, reduced cost and reduced cycle times for all participants in the extended enterprise. The goal is to provide enablers for achieving cost savings, lead-time reduction, and quality improvement in the global automotive supply chain through collaborative efforts involving the use of vehicle product data. This vision however, is a long way away from our current situation within the automotive industry.

2 Changes Required for Collaborative Engineering

Collaborative Engineering as practiced today does not bring about the above mentioned benefits. Development is only part oriented rather than system oriented and data exchange is carried out using privately developed interfaces. Each party has their own processes and immense coordination is required to manage the various interfaces, processes and data sources. This means therefore that similar solutions are often unnecessarily developed by various parties. For each new collaboration new solutions are often developed leading to huge amounts of redundant data and a waste of resources. A huge amount of personnel is required to coordinate and organize everything, especially late in the development phase.

2.1 Change of Roles in Development Processes

An OEM, for example, works with several suppliers and, in the worst case, an individual interface is required to each supplier so that data can be shared with this particular supplier. This of course leads to a huge amount of data being transferred to and from the OEM. The supplier and OEM can not normally access each others systems so the interfaces normally lead to an external database which can be accessed by both parties. The OEM has to ensure that all the interfaces to the supplier are running quickly and smoothly and the supplier has to ensure they are receiving the correct information on time from the OEM. Once the supplier receives the information they have to start their own internal process. At certain points the supplier will have to return information back to the OEM and this is also done through an interface. The OEM also has their own internal process and this has to be fed with information from the supplier. Both the supplier and OEM therefore have to contend with internal and external processes and a large number of resources are required to organize all this. It is easy to image the complexity of this on a larger scale. One OEM for example could have more then 15 different suppliers, each requiring their own interfaces and databases (see figure 1).

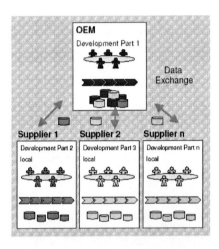

Figure 1: Collaborative Engineering today

This problem could be solved through the use of Collaborative Engineering and globally organized teams. With Collaborative Engineering development is system oriented rather than part oriented. Tasks are globally organized and are partner overlapping so that each partner does not need to carry out tasks that can be carried out centrally. Through globally organized teams each partner can save on resources or reassign people to carry out other tasks. By connecting processes and systems, data can flow easily between each partner meaning less data is redundant and the coordination of the various processes and systems is minimized. Development tasks themselves can also be easily shared between the collaboration partners as each person is able to access shared systems and data. This means less resources are required across the partners as one clearly defined person is responsible for certain activities across all partners, rather than one person in each partner company being responsible for certain activities within their company (see figure 2).

This makes one of the most important factors in the automotive industry possible, the use of platform strategies. Platform strategies allow different brands or companies to use one platform (the core technology) to produce vehicles, which can then be adapted for the local market. By reusing this core technology partners can save huge amounts of money and time. Adaptations for the local market however still need to be carried out as not all vehicles can be sold in all countries.

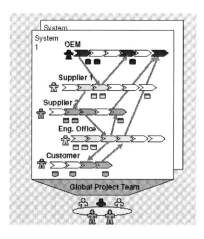

Figure 2: Collaborative Engineering in the future

2.2 PDM Collaboration

As mentioned above, one of the major requirements for Collaborative Engineering is that data and systems are integrated (see figure 3). Users need to be able to access data from several systems, almost certainly from different continents and from different partners. As data needs to be available online 24x7 a heterogeneous partner-overlapping system is required. One of the most important things required before this can happen is the definition and creation of standards, which describe the data and functions. Also required is the definition of a framework for this online collaboration. An authorization and user access rights concept is clearly very important due to the number of users accessing various systems and data across companies. What however is not required, is that a certain user has to log on to each of the systems each time they want to access data in another system or at a partner company – once they are in the system they should be able to access all data that they are authorized to access without having to log on again.

Such a collaboration framework was developed by Daimler Chrysler Research.

The function of the framework is to enable flexible connection of different engineering partner systems using web services and to enable online collaboration between the systems. Therefore the development of system adapters was required so that system functions are available externally. A connection to the native systems is usually possible by using the system specific Application Programming Interface API. The adapter undertakes the required mapping between the native representation and the formally described representation of the system specific functions. This allows the most important functions of the systems to be externally available for read and write access.

This approach provides an enormous advantage. Heterogeneous systems with different functionality, different system design and different system philosophy regarding data management are easily and flexibly connected. Of great importance in this case is the fact that the system interfaces, adapters and the required Web Services are built up similarly or are standardized. The required standard for this, the PLM Services, is on the way to becoming an international OMG Standard in the next few months [2]. This is detailed further in the next chapter. A technology for the management of the partner overlapping work share is required to realize Collaborative Engineering in a partner network as well as the development of a solution to connect the partner systems. This is accomplished by an especially aligned Engineering Framework.

The prototype Framework, Common Core, is split into four models (see figure 3).

Model 1: Description of the partner overlapping data and the relations between the data. This includes data for configuration management and versioning.

Model 2: Description of the partner overlapping processes including data from change management (initialization and steering of change processes, release process).

Model 3: Communication Scenarios (Use Cases), which are to be used from all partners. One example is the partner overlapping request for change.

Model 4: All data to configure the partner overlapping Collaboration (user rights, authorities, role concepts etc.).

Based on these data models there is a set of various collaboration services available for all partners. This set of services can be extended as they are flexible and it is possible to develop special services for particular requirements. Fields for special services are, for example, the management of bill of material data, shared parts, to integrate design and production data. Important for all expansions of the service set is that all changes regarding the models as well as changes regarding the web services are aligned to the standard PLM services[2]. Thus, a powerful tool is available which enables the building up of an Engineering Framework, designed by single flexible and web based components, which affords a flexible integration of different partners with their heterogeneous system landscape and which fulfills the requirements of strategic development partnerships (Value Networks) in an optimal way. By integration of additional tools such as virtual conferencing, 2D/3D viewing, project management etc. it is possible to buildup a very capable solution to manage engineering cooperation. Because the communication between different frameworks is based on the same technology, an integration can be easy realized if necessary.

The implementation of standardized collaboration platforms as described above will enable the reorganization of the work share between the OEM and their supplier for the first time towards more design responsibility and will considerably increase the efficiency in design.

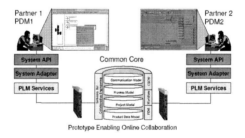

Figure 3: Integrated data and systems

2.3 Process Standards

The integration of systems and data however only fulfills part of the requirements for Collaborative Engineering. To gain the most from Collaborative Engineering, partners need to integrate their processes. If this is not done data becomes redundant and the benefits of Collaborative Engineering are not seen. If each partner, for example, has their own change management process, data about the changes needs to be fed into two different processes. Managing and coordinating these changes becomes complicated and, in the worst case, may become impossible. Change Management is of course not the only process, which should be unified. Release and problem management become much easier if one process covers all partners. Not only is the process easier to handle if it is the same for all parties, but communication is faster and less travel is required as collaboration is easier and less face-to-face meetings are required. In addition, data handling is at a minimum and fewer errors are made so one can be sure that data at both parties is the same. For this to happen, both parties need to use the process and systems locally but access to relevant data is given through a view for external parties (see figure 4).

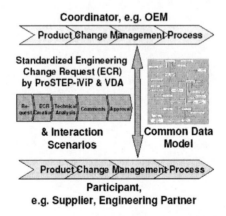

Figure 4: Integrated processes

Especially for the depicted scenario, where both parties have their own change management process and system, the heterogeneity of the standardization effort needs to provide a way to overcome their heterogeneity. Therefore, the Engineering Change Management Working Group [3] developed the approach to use a harmonized change management request process for communication purposes (see Fig. 5). Thus, if an interaction with the partner occurs, the coordinator party leading the change and participants that are involved, map the view on their internal change request process to this harmonized process. Based on this common process view, messages may be exchanged between the parties at certain synchronization points that have been identified in the harmonized process. Interaction scenarios provide common message patterns that are typical for specific engineering cooperation models such as a module supplier or engineering partner [4]. For the data contained in exchanged messages, a common data model was developed based on the existing internal data models. This data model describes the requirements for data exchange in order to communicate using STEP AP 214[5].

This standardization effort was carried out with high attention, especially as it is an important pioneering piece of work in the engineering field and the process is very important. The potential for the expected benefit is quite high when considering the large volume of several hundreds of change requests exchanged in some cooperation projects. The working group was able to specify the process, the interaction scenarios, and the data model without many general discussions, even though the internal change management processes and systems are quiet different. This was possible due to the unity among the participants and the methodology to start work by identifying common use cases, to identify the common process and to use that as a common paradigm for the development of the further results.

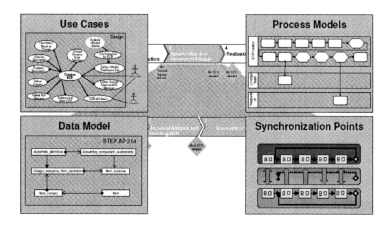

Figure 5: Standards for Engineering Change Management

2.4 WEB Standards

Due to the various data formats and systems, standards are required to define data and functions and to ensure that all parties understand each other. One new upcoming Standard for Collaborative Engineering is PLM Services [2].

The standards currently available only cover a part of the collaborative engineering requirements. Engineering collaboration as we know it today only allows asynchronous, file-based data exchange (data exchange in batch mode). The STEP/part 21 standard (physical file) is very often used to describe the data to be exchanged in a neutral format. PLM Services is the first standard, that not only describes the data required, but also describes the functionality needed for collaboration (see figure 6) and thus also allows synchronous data access (online access). It is this synchronous data access that provides the basis for implementing realistic engineering collaboration scenarios with a combination of online access to the partner systems and the simultaneous down or uploading of engineering data with the file-based, asynchronous data exchange.

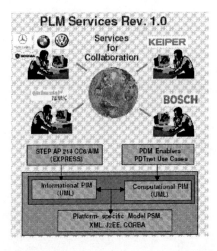

Figure 6: PLM-Services, a new standard to enable Collaboration

The realization of the standard took OMG's Model Driven Architecture (MDA) into consideration [6]. The basis of this standard is a Platform Independent Model (PIM), divided into an information model, which describes the data and a function model (computational model) which describes the functional aspects (use cases) needed to run online collaboration. The PIM information model is derived from the STEP AP214 CC8 (Conformance Class eight) Application Interpreted Model (AIM). Both models are linked by an EXPRESS-X mapping which is part of the standard. Through this a direct data-related link between the ISO model and the OMG model is established.

The functional model defines the basic functions in the form of use cases, a description of interactions between systems in the case of collaboration. The method to describe the use cases is defined in the PDTnet [7] and the PDM Enablers[8] project. Examples for use cases are:

- authentication of access to a system
- a generic query and traversal mechanism
- functions for reading and writing information
- functions for up and downloading in synchronous and asynchronous scenarios

This function model (Computational Model) and the data model (informational Model) are aligned with each other. Together, the two models provide the basis for direct IT support in system-to-system collaboration. The Platform Specific Model PSM describes additional information, which is required to buildup the platform specific implementations based on XML, CORBA or J2EE. The functions are described in the W3C (World Wide Web Community) conform Web Services using WSDL (Web Service Description Language) and SOAP (Simple Object Access Protocol) as the protocol for the specified queries. The native data (e.g. CAD files) will be transported as STEP part 21 files. Thus, a fully web-based approach based on ISO/OMG/W3C standards is available to describe all aspects of realistic Collaborative Engineering scenarios. It is also the first standard to bridge ISO<->OMG<->W3C. The standardization of the PLM Services is an important milestone on the road to enable the long-heralded vision of engineering collaboration or for E2E to become a reality. PLM services was created by a consortium of automotive suppliers, OEMs, IT suppliers, members of the research community and standardization groups [Produktdatenjournal]. The project started in January 2003 and a reference implementation is currently being created. The Standard was accepted from the OMG Architectural Board in April 2004 by a formal recommendation to the OMG Domain Technical Committee (DTC), which means that the path to standardization of the PLM Services 1.0 is clear. It is planed to extend the standard to Engineering Change Management requirements (PLM Services 1.1).

2.5 Psychological Aspects

Collaborative Engineering of course does not just rely on the integration of data and systems, it relies on the collaboration of people. Despite their potential, most diverse teams either do not perform as well as homogenous teams do, or they equalize [9]. Good management as well as good collaboration methods and tools makes the difference. In the past, distributed collaboration could only be supported by telephone or business trips. The difficulty with telephone conversation is that it is impossible to see the other person as well as to see the documents they are looking at. Telephone conferences are also difficult if people have to speak a foreign language. Business trips and face-to-face meetings are a better alternative, however, they are very expensive, time consuming and can not always be arranged at short notice. Today, video conferencing sometimes is possible but is very expensive and the video-conference rooms are not always available. Even though the partners can see each other, the social awareness is restricted, because of missing eye-contact. Furthermore it is not possible to view the partners computer screen or documents they hold in their hands. Collaborative Engineering requires more than this and virtual conferencing takes video conferencing one step further. With virtual conferencing each person has a camera at their workplace and can use this to participate in virtual conferences. This means people can easily collaborate with others without even having to leave their desk. Virtual conferencing is web-based and is therefore a low cost solution. As well as being able to view the other person, computer screens can be viewed and documents can be shared. This is a very simple but effective way of working together and helps solve some of the work satisfaction problems that were mentioned earlier. Essential for the success of collaborative engineering is sound communication and harmonized teamwork between various "intercultural" partners. Not only does a common language have to be defined so that parties can communicate with each other, but common project terms need to be defined so that it is clear what is being referred to and it is certain that all parties have the same understanding. As already mentioned, a role and authorization concept is required for access to shared systems and data to ensure team members can access the same information. The rules for these access rights need to be defined so that each partner can assign rights in the same way. In addition, each team member needs to be trained to use the processes and systems, so a common qualification program should be developed. A project language and glossary of terms also helps team members communicate with each other by providing a standardized overview of common terms and words. In many cases these standards are not fully defined or do not exist which leads to miscommunication within teams or partners carrying out processes differently. In the worst case, misunderstandings can lead to members of the same team working on different solutions, which may not be able to combined at a later date. Again, this results in work satisfaction problems for the team members. As well as a general project language or glossary of terms, different systems and methods may need to be mapped to each other. Fields in one system may hold the same information as in the other system but the label may be different. The systems themselves do not necessarily need to be adjusted as long as the differences are clearly documented and available to the project members. Taken one step further, one part may have two different part numbers at different companies. It would be a very complicated and time consuming process to create one new part number for the part so a cross reference table could be developed so it can be identified by both parties. It is not always necessary for both parties to use identical systems and processes – in fact it is important, that each member is aware of the differences (see figure 7). While it is important to have common standards, methods, processes and languages it is important to remember that local identity and culture is very important [10]. The approach in this context is to find an appropriate balance between the local identities of different sub-companies and the global identity of the main holding. Collaborative Engineering does not attempt to standardize people so that they all think and act in the same way. The local language as well as the local context for example are important while working with colleagues that are from the same area/region. People feel comfortable speaking their own language/dialect and this should not be stopped. Only when a person they are working with does not understand the local language should a common language be used. In addition local laws and policies are also very important and again they should not be ignored. These laws and policies can not be the same across the globe and in some cases products may need to be developed differently to meet local regulations. Failure to adhere to these standards would be a catastrophe for the company. Consumers and their local preferences also need to be taken into account. These preferences differ between people in different geographic locations. Many consumers in America for example like large cars whereas consumers in Asia prefer smaller cars. The reasons for this can be found in the road network at the relevant countries. The roads in the

USA are generally rather large and parking lots provide large parking spaces. Streets in Asia however are generally smaller and more crowded so a large car would be inconvenient.

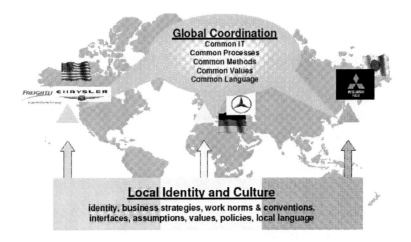

Figure 7: Intercultural cooperation

The appropriate balance between globalization and localization is an essential factor for Collaborative Engineering. A certain amount of change is required for the success of Collaborative Engineering but a certain amount of local adaptation is also essential. Experiences show that intercultural engineering becomes efficient if the participants are motivated, they have common goals, their identity is not questioned, they have interdependent tasks and they trust that the project can be a success. However, one of the most important factors is the acceptance of the decision process. One of the best ways this can be done is to combine the best practice of all parties so that each party is able to recognize their input.

3 Conclusion

Briefly summarized, we have seen that Collaborative Engineering is the most effective way for automotive companies to increase economic growth. It will create new partner alliances, which can transform the way engineering is carried out in the future. However, finding the correct balance between standardization and localization, as well as the motivation of people to change their working habits are the major challenges are facing companies. It will therefore require a joint effort between all parties (OEMs, suppliers, IT suppliers, research institutes and standardization groups) to ensure success. Collaborative Engineering will change work patterns and will do pioneer work in intercultural cooperation. It will bring about major benefits for companies within the automotive industry, and if applied correctly, will ensure success for those involved.

4 References

[1] Ram D. Sriram, "Distributed Integrated Collaborative Engineering Design", Sarven Publishers, Gaithersburg, USA, 2002
[2] Michael Feltes, "XPDI – Standardisierung von web-basierten PLM-Services für das kollaborative Engineering", ProduktDaten Journal, Nr.1, 2004
[3] ProSTEP iViP Working Group "Engineering Change Management" (ECM) http://prostep.org/de/arbeitsgruppen/ecm/
[4] VDA AG Kooperationsmodelle – Datenlogistik. "Handbuch Kooperationsmodelle Datenlogistik.", 2001
[5] ISO 10303-214, "Core Data for Automotive Mechanical Design Processes", 2003
[6] OMG,"Product Lifecycle Management Services V1.0, Revised Submission", http://www.omg.org/mantis, 2004
[7] ProSTEP iViP PDTnet Project, "PDTnet-Project Documentation", http://www.pdtnet.org, 2000
[8] OMG, "Product Data Management Enablers V2.0 final submission", http://www.omg.org, 2002
[9] Manznewski, M.L. & DiStephano, J.J., Global leaders are team players: Developing global leaders through membership on global teams. Human Resource Management, 39, P. 195-208, 2000
[10] Grant, E., Schulze, H., Haasis, S., Intercultural virtual cooperation: Psychological challenges for coordination, Conference proceedings of Human Computer Interaction International 2003.

5 Authors

A. Katzenbach, DaimlerChrysler AG, Wilhelm-Runge-Str. 11, 89081 Ulm,

Alfred.Katzenbach@daimlerchrysler.com

M. Feltes, DaimlerChrysler AG, Wilhelm-Runge-Str. 11, 89081 Ulm,

Michael.Feltes@daimlerchrysler.com

J. Willis, Protics GmbH, 628-P305, 70546 Stuttgart,

Joanne.Willis@daimlerchrysler.com

Engineering Change Management as a Part of Electronic RFQ

Rolf Klamann

Continental Teves, Frankfurt, Germany

Abstract: The described new process brings CAS a decisive competition progress in communication with their suppliers. The fast information flow by using WEB technology in a very secure environment is a milestone for CAS and all suppliers to get much more benefits in the future.
But the evolution will not stand still: CAS and Supply On is preparing more efficient solutions especially in the area of project purchasing in a very early product development phase. In 2005 there are scheduled some pilot projects to get experience with WEB-based collaboration tools and methods. This will be another new investment to integrate the future in the Automotive Industry.

1 Introduction

The ability to innovation, flexibility and the potential, to react rapidly to individual requests of the automotive market, are criteria for competiveness of an enterprise in the global competition today.
The worldwide globalization in the Automotive Industry needs more conditions for effective cooperative work between OEM and their suppliers. The permanent reduction of product development time in parallel demands a drastic increase in process quality and process security especially on the suppliers side.
Large companies have today two alternatives to design electronic purchasing processes via WEB-technology: They can built up a company-specific portal solution with a very complex IT infrastructure. Alternatively they can map the requested functionality by connection to an Electronic Market place.
This concept has been picked up in 2000 by Bosch, Continental, INA and ZF.

They decided to establish the Electronic Market Place Supply On as joint venture of global automotive suppliers.

2 The Electronic Market Place Supply On

The concept of Supply On is a communication platform from suppliers for suppliers.

Supply On is more than a commercial market place for electronic purchasing and auction biddings, but has additional process functionality in the area of Engineering Collaboration, Supply Chain and Quality Management.

The characteristics of Supply On Sourcing are the Business Directory, the Bidding Module and the Catalog Solution. The Business Directory offers more than 100 material groups for quotation requests by purchase departments.

Bidding functionality and catalog parts complete the offer of Supply On Sourcing.

The Engineering part of Supply On contains the Document Manager, a collaboration module (Project Folders) and a link for CAD data conversion.

The main functionality is the Document Manager, which handles internal standards, drawings, specifications etc. of the different shareholders of Supply On in a very secure environment for a closer communication with their suppliers.

The Supply Chain Management structure of Supply On deals with order management by WEB EDI, Vendor managed inventory, management of empties and additional logistic services.

Two modules characterize the Quality Management offer of Supply On. The Performance Monitor allows the online quality control of the supplier production (e.g. ppm-rate).

The Problem Solver supports the creation and management of 8D-reports to the request of the OEMs.

3 Quantum Leap in Process Communication

An improved Engineering Change Management (ECM) process has a significant impact on companies´ capabilities and effectiveness. This fact was the reason for the request of the purchase departments to have a direct link between Supply On DMS and Sourcing together with a backend integration to the inhouse PDM (Product Data Management) System at the Supply On shareholder Continental (in the area of Continental Automotive Systems).

Engineering Change Management as a Part of Electronic RFQ

The link between SAP/PLM and Supply On DMS has been realized at Continental Automotive Systems (CAS).

The target was to give the CAS suppliers access to the latest released technical documents, not only during the RFQ phase but also after start of production including Engineering Change Management.

Each RFQ is accompanied by different technical information like drawings, bill of material, standards, specifications etc.

In the past the purchase department had to order these documents in paper and to send them to the suppliers. Because of security requirements the documents were not allowed to send as attachment by email.

Inside Supply On procurement it was possible to create those documents as attachments in a secure 128-bit code environment. The disadvantage was that each single document has to be specified in SAP-PLM manually transferred to Supply On.

In many cases these 'preparing' process took 2 or 3 weeks until all information had been completed on the CAS supplier side.

At the background that currently the reduced development time of a new car platform takes only 24 – 30 months this period of 3 weeks during the RFQ phase is much too long.

A second hindrance in the past was that the suppliers were not integrated in the internal Engineering Change process. The Engineering Change information has to be transferred to the supplier by the people of CAS purchase department. Business trips or holidays of the specific person in the purchase department result many delays in information transfer by the push principle in the past.

That means in worst case that the supplier had no or too late information about Engineering changes of parts, material, specification etc. This could have impact to the current production so that the supplier produced 'wrong' parts.

Since mid of 2004 the RFQ process at CAS has been completely changed.

With the increase of WEB-technology the possibility of communication have much more flexibility in worldwide operation of the automotive industry.

The new RFQ process at CAS describes a quantum leap in the communication with the CAS suppliers, but also in the internal process inside CAS.

CAS built up the so called 'purchaser workbench ' inside SAP-PLM. This user interface permits the purchase department the access to document packages in contrast to single documents in the past. A document package contains all released

49

drawings, bill of material, specifications and standards identified by the specific material number. This functionality facilitates the RFQ purchase process considerably.

In contrast to the former process the purchase department don't transfer documents but only XML-files with the metadata from SAP-PLM like part number, part description, release level etc. The supplier must play now the active role to get the access to the documents like TIFF or PDF-Files. The process changed from push to pull principle. The supplier gets the access to the documents via metadata and specific access rights which will be released by the CAS purchase department. Then a document download process of TIFF and PDF files can be started by the supplier.

A great challenge at Supply On side was the encoded environment, which is absolutely necessary for the transfer of sensitive data like technical drawings especially in the very early product development phase of the Automotive Industry. The security officers auf the Supply On shareholders have specified the data security demands to be realized by Supply On.

Therefore Supply On as the first electronic market place has been certified according BS 7799-2 (British Standard of Security). This was the strongest requirement which has been fulfilled by Supply On as assumption for using the Document Manager.

4 Author

Rolf Klamann

Continental-Teves AG & Co oHG, Guerickestrasse 7, 60488 Frankfurt,

rolf.klamann@contiteves.com

Generative Car & e-Vehicle

Peter Kraus

IBM Germany Product Lifecycle Management (PLM) Solutions
PLM Consulting – Automotive, München, Germany

Abstract:

Product Lifecycle Management is one of the most emerging markets worldwide. All mass producing companies face this PLM challenge, although the scope is defined quiet heterogeneous yet. Meanwhile the PLM framework can be considered to be placed in the market and also accepted. But the PLM definitions still variegate and are not consolidated. So one of the most innovative industries - the German automotive industry – is approaching this PLM challenge in two main directions:
One is the associated parametric design approach. In a vision, this may comprehend a complete car or even car series:
→ IBM focuses this future with "The Generative Car".
The second is the penetration of the vehicles with electronic and SW applications with its extreme short development cycles:
→ IBM focuses this future step with "The e-Vehicle".

1 Introduction

The PLM business is a quit complex one. To structure it is a pre-requisite for not being lost. So it is helpful to differentiate two main areas. One is the "traditional" design work, while second the management of engineering data is getting one of the current success factors. Each of this area needs to be handled as separate program with a strong interface to each other. And within each of these programs, there are so many facets that each of them should be handled as an own project. This PLM project structure will help to serve the "real project": The product development of cars or even car series.

More then a decade ago it was helpful to get a design tool, where we came from drawing to modelling. At this point it was necessary to bring together engineering knowledge with tool know how to get the design done. Then we fit together some parts - this we called DMU (Digital Mock Up) – to find out whether the number of modelled 3D parts still fit together in an assembly. So the referencing between parts and modules – not yet geometrically associated - became an additional need in the engineering word; This was most often handled by manual prepared tables. The ERP systems could not take over this task.

A huge side effect of this PLM approach was the usage of the same product information in product design related disciplines like tooling, viewing facilities i.e. for product maintenance, and many other. To manage this unlimited interoperable usage made an "engineering PDM" necessary. This brought up another pre-requisite: The controlled knowledge of data flow within the product development process (PDP). The engineers needed a lot more of process knowledge and how their design works were interlinked in the PDP. In our days the engineering PDM is already in implementation mode. In the nearest future, we are forced to find ways, how the "Contextual Design" can be handled. "Contextual Design" means that the parts, which are design in context into a "Product Structure", will be enhanced with parametrical associative links.

The vision will be to enhance, managed and maintain these geometrically linked parts, assemblies, devices, including electronic devices across series. Then the Contextual Design for a complete product, the "Generative Car" will become reality.

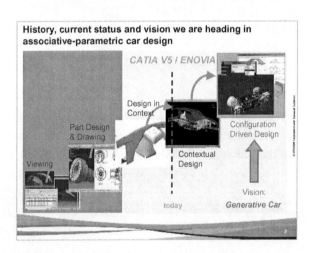

Figure 1: Generative Car

Generative Car & e-Vehicle

The second challenge is the electronic, or let's be more precise, the SW application within a car: Currently more the 40% of innovations are based on electronic features in Premium Class Cars – and it will increase up to 60% in 2010. Until today OEMs handled these innovations individually in fact as their own Individual Property (IP). This resulted in a network of up to 700 functions managed by up to 80 independent processors and their operating systems. The electronic complexity is exploding and needs also to be restructured. And as these electronic systems and functions have a quiet shorter development cycle then the mechanical car, two different development cycles have to be brought together. This is a real new challenge.

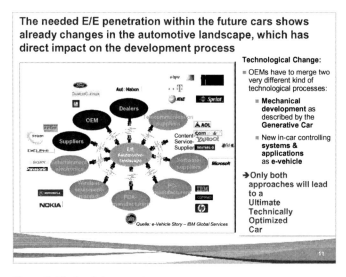

Figure 2: Electronic increase in cars

To solve this conflict the restructuring focuses also on two main areas: On one hand it seems to make sense to separate basic IT infrastructure in a car and to compete in functions/ SW applications on the other hand. An individual car IT infrastructure, also called system architecture does not demonstrate progress, innovation or emotional binding to the product by the end user – the drivers. So there is no need of an OEM to do it for his own for IP differentiation and in consequence it can be commonly designed by OEMs. And in fact, this is a realistic and currently initiated approach in the market, so called AUTOSAR (see Internet). As this approach is to new to be considered in the product lifecycle management, it will not be discussed in here.

53

What seems to be more interesting for Systems Supplier is the fact, that the numbers of extreme accelerating SW applications - which will make the real difference between the cars - must also be developed, managed and maintained.

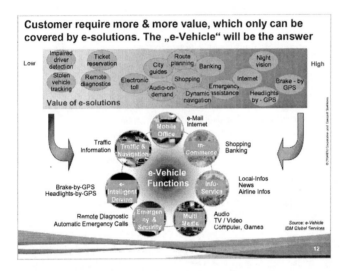

Figure 3: e-Vehicle

A management system must be put in place to handle these applications, which represent the resulted functions. ➔ The "e-Vehicle" needs to be managed by an "Application Lifecycle Management (ALM)". At least, this SW application management must be linked to the electronic devices, which are an integral part of the traditional PLM business.

2 Generative Car

2.1 Roadmap to Generative Car

As mentioned in the introduction the PLM systems must support the PDP. So fundamental process knowledge must be linked to the design process. To optimize the engineering system management will not be enough on the way to the "Generative Car" because it will only optimize the design components and it will not optimize the (complete) PDP. In consequence OEMs, with their design and process knowledge and systems suppliers, which must also have considerable experience in process consulting, must closely work together – each with their best practice experience. The framework of this working together will be the "Digital Reference". It is the kernel in the process as it is the "real product" as it gains maturity – even it is "available" only virtually. At least it can be modelled, visualised, optimized and released. The future extent of the product validation process might include more and more parts, modules and functions, until "everything" is confirmed virtually.

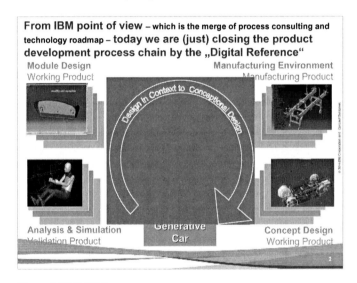

Figure 4: Digital Reference

The reason, why the "Generative Car" will become true is for sure again economically. Efforts to validate each single product will not be possible any more. Results of a concluded validation from one product must and will be transferred to other ones. The market demand to deliver more and more verities of cars will also not automatically lead to an increase of engineering power as the pressure on costs will continue. So the deployment of engineers will - in best case - have parallel growth as the market share of an OEM. ➔ In consequence the enhanced complexity of products (= cars), the enhanced numbers of product verities (=car variants), which must be served with almost the same number of engineers will lead to an extreme support of all kind of linked relations. This complexity in Contextual Design can and will only be solved by an evolutional introduction of heavy design and data management systems, which will in turn lead to adapted PDPs and further organizational paradigm shift.

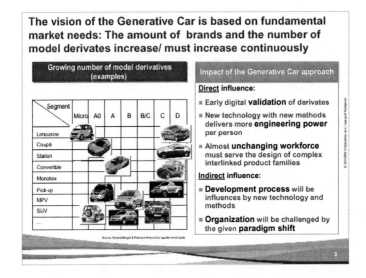

Figure 5: The market drives the Generative Car

2.2 PLM & Generative Car

As mentioned above, the extremes of needed innovations within car development on one side and the cost keeping on the other side will force the OEMs to manage their products with heavy CAx and engineering PDM systems. IBM as the biggest system supplier and consulting company in one hand will have the power to support these extremes by merging and leveraging this knowledge in its organisation to bring it to the market. But the complexity will continue to increase exponentially – just consider reuse efforts, innovation pressure and aesthetics performance (so called design study). On the way to the 'Generative Car' the mentioned pre-requisites should be solved by a preceding discussion on proposed ideas and concepts. This we call 'Thought Leadership', which can be delivered by "consulting system suppliers". This Thought Leadership has primarily to consider a) in which way the industry is heading in the future b) define and deliver appropriate technology c) deliver mature technology and/ or set the right expectations on development speed on the "heavy PLM systems" and at least d) PLM implementation speed.

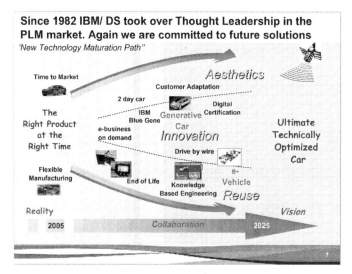

Figure 6: Thought Leadership on PLM means PLM Consulting with PLM systems power

Another framework PLM has to consider is the balancing of four main success factors
1) Development Knowledge,
2) Product Development Process (PDP) optimization,
3) Technology enhancements and
4) People Organization

We will not succeed, if one of these success factors is quite good in place, while one of the other will suffer in maturity. And there is no metrics in place of what is balanced. This in turn makes it again difficult to give proven recommendations on needed investments. Even for "PDP consulting systems support", which really can bring in the experience from previous projects it is hard to disclose a benefit matrix. But currently there a several considerations which help to find the right focus: a) It is the accredit experience on best PLM practice models, b) demonstration of the technical roadmap for the previous mentioned PLM s foci, c) knowledge on balanced success factors, d) experience on the knowledge flow within the PDP (= success factor 2).

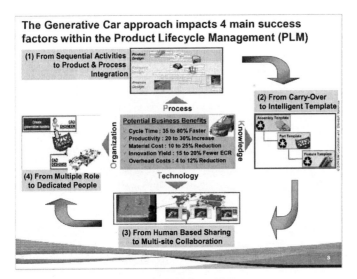

Figure 7: Four success factors in PLM Consulting

Knowledge is more then the management of design and processes. It is the controlled reuse, share and administration across the complete PDP. Complete in this sense means, we consider the knowledge flow across company borders and it takes place in an end to end PLM view with commonly share rules: → This will be a real "Extended Enterprise".

IBM as system supplier and consultancy company in one faces the challenge to understand this knowledge flow at our customers and provide best PLM advice. We will take responsibility through PLM 'Thought Leadership' and by delivering best practice ideas and concepts, supported at least with knowledge based systems!

Generative Car & e-Vehicle

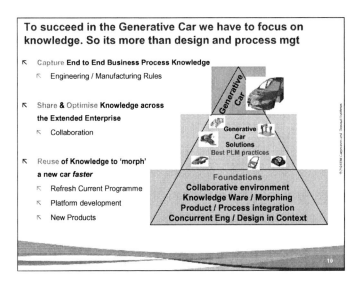

Figure 8: PLM Knowledge to succeed the Generative Car

3 e-Vehicle

As mentioned in the introduction, the functionality of the "e-Vehicle" will have at least two main SW layers. The one, which is normally neither interesting nor transparent for the end-user, the driver, and the one where he "gets what he wants". This situation, on an abstract layer, can be compared with a personal computer.

The inside of the computer, like processor, networking or processing of the required data (text, numbers or pictures) is not what the end-user wants to handle. In this infrastructure level the requested applications are processed and managed. This operating environment will not be considered in here.

What the end-users (we) want to handle are the functions as text- or calculation- programs or clear moving pictures etc. Here is an important relation: Required functions are written by applications or application programs (functions directly relate to applications). This logic can be assigned on the behaviour of a car: The driver is interested in functions like Route Planning, Electronic Toll, Brake by GPS (which had identified a dangerous situation like an accident or road constructions ...) and many many more ideas of what can be done in future. These functions will increase irresistible and all of them must fit to the system architecture of the cars. Therefore the

AUTOSAR cooperation will define very few – ideally only one – Application Interfaces (API) on which these number of application will be fit in.

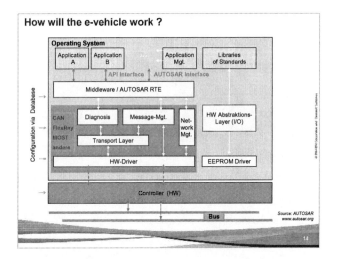

Figure 9: PLM Consulting and Systems are one key success factor

Why is it important to support this OEM approach? It is, because OEMs must manage a new business. The request from the market to get value hasn't changed. So the OEMs will bring innovations to the end-users by offering solutions. But the point is that these solutions most often can not be performed by mechanical design as it had been possible over 100 years. They have to enter the SW market. And this market has different rules – most important the different cycle time, diversity and its release and maintenance management. The "layer", where these differences are handled are on the function level. Here the solutions are performed by a verity of functions, which interlock, are cascaded, work in loops and ... and ... and. The management of these functions, better to say applications, which perform the function, need the same logic of a lifecycle management as mechanical parts do. So we are entering a synonymous to the PLM. I will call it ALM (Application Lifecycle Management).

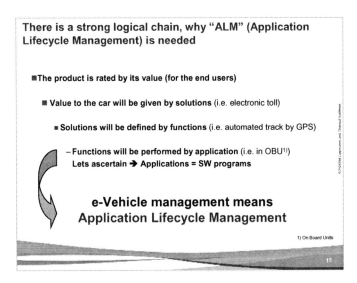

Figure 10: ALM (Application Lifecycle Management)

4 References

[1] IBM Internal Knowledge Database: ICM (Intellectual Capital Management - Automotive
[2] AUTOSAR **Automotive Open System Architecture** http://www.autosar.org/

5 Author

Peter Kraus, IBM Germany, Hollerithstraße 1, 81829 München

peter.kraus@de.ibm.com

Nissan digital process and next steps

Daisuke Shimomura

GLOBAL IS DIVISION SYSTEM DEVELOPMENT DEPARTMENT, Nissan Motor Co., Ltd., Atsugi-city Kanagawa, Japan

Abstract: Nissan has been using an accelerated 3D data centric development process since 2000 when we started to make real use of a 3D solid based CAD system. This system provides precise digital mockup, CAE numerical analysis, productivity study, paperless drawing, etc. 3D data was enlarged year by year and distributed to many development offices. I attached the benefit of 3D digitalization and forecasted the key issue in near future.

1 Introduction

Middle of 1990's, we started to reduce the development period. Now, we have 19 month development period. At first PDC **Data quality assurance** (we call the prototype car development process "Plan Do Check") process, we can feed back to production release by prototype test. But now, we don't have any time to feedback to production release by prototype test result. We have to enrich the digital engineering period. Therefore the prototype test was replaced by DMDR and digital validation. We started full digitalization development way through 3D digital data stream.

2 Current Nissan 3D data centric process

2.1 IT activity within development process

This is a figure of overlapping work in each division. After the planning process, styling, engineering, prototype production, test, product engineering and sales preparation work concurrently. In each division, now, the digital data is used a lot.

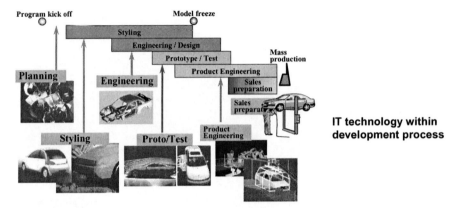

Figure 1: IT technology within development process

At the Design styling phase:

Detailed design reviews ranging from lens-cutting of lights to total coordination, the screen-expression technologies including backgrounds and motion pictures, and the one-on-one full-scale projector, all of them have brought about the shorter-time design reviews and examinations with a true-to-life sense.

We are now trying the Perceived Quality improvement as one way of production quality enhancement. Until this point, we had to rely on the feedback from a proto-type car only to evaluate total impression for the kind of combination of different materials, but now CG has taken its place.

At the Engineering and product preparing phase:

Because of the core technology of the digital mock-up, we have seen a huge change by the application of the Solid technology. The three-dimension CAD has drastically improved design review precision including production requirements even in the design planning stages, as well as optimized data use for analysis, CAM and production preparation.

By the Digital engineering study, the digital data can deal with the entire process that the conventional trial and experimental phase handled. The point of Digital engineering study is that developers share the three-dimensional digital data and simultaneously conduct part examination, layout review, and productivity checking with consistency.

CAE's one of the most contributing factors in development time and cost reduction, as well as in performance enhancement, is the crash simulation. CAE has shown overwhelming achievement by drastically reducing the number of proto-type cars and the experimental period.

Product engineering study can also be done at early stages of development. We have almost no opportunity to do some physical evaluation before the production release.

Sales preparation phase:

The impact of a shorter development period is not limited to the R&D area, but also brings about the shorter sales preparation period including catalogs and service-manual creation. The real vehicle and its illustrations are the conventional way to create catalogs, but recently the use of CG has become rapidly increasing in view of time and cost reduction.

2.2 Paperless drawing system applied in 2001

A paperless drawing system was launched in 2001 as real use. Nissan design engineers concentrated their responsibility to 3D data building, then we reduced the responsibility to make 2D drawings. We finished the delivery of paper drawing. The 2D drawing was divided in three parts. The first is 3D CAD data with annotations. The second is supplement drawing if necessary. That is written engineering key measurement and sections. The third is data note that described the history of release. These files are sent to suppliers and manufacturing plant with BOM data by electronic data, so there is no paper delivery any more.

2.3 3D data utilization in plant

In the plant, the people don't need 3D CAD data itself. Because, they don't have any needs to modify 3D CAD data. So we prepared a viewing system called "Space Vision" .It is an in-house program, which is also sold to suppliers. We developed the specific functionality for the Nissan paper less drawing system, then, engineers in plant can easily operate the viewer system and the digitalization in plant office was started.

3 Benefit of digitalization

The second half of 1990's, we have challenged a full digital mockup design review on TINO program. In that time, we have developed TINO within 15 month. At TINO program, we created surface data for digital mockup as additional work. Because, at that period, we had used in house wire frame based CAD system. After this success, we launched I-DEAS solid CAD system and applied this digital process to all vehicle development programs.

This figure is a number of design changes. We got big reduction of design change by digital process. Nissan estimates that 50% of the contribution of design change reduction is coming from digitalization.

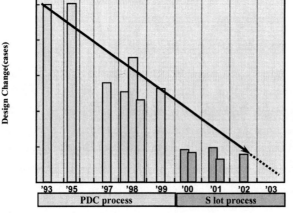

Figure 2: Reduction of design change

Following the digital shift, we got a certain number of reductions of prototype cars. The model revise cost was also reduced by the deduction of design change.
Finally, the total development cost was reduced roughly 30%.

4 What's a next step?

4.1 Global operation by digital data

Now, business environment is changing faster than in the past. The first, concurrent vehicle development was started within worldwide development offices. For example, the platform is developed in Japan, and at the same time, the upper body is developed in the Detroit branch. Japan and the office in the US work concurrently through digital mockup. Sometimes, Japan chooses US suppliers and the US suppliers check in 3D data in the Nissan Japan database for DMDR. The real time corroboration is mandatory worldwide. The second, the Alliance was started in 1999 with Renault. We have planned a common platform development and a new engine development and so on. Not only 3D CAD data exchange but also precise attributes of parts is necessary for data sharing now. Nissan also started alliance with Chinese companies too. The alliance activity will expand year by year and we need to prepare global infrastructure. The third is data quality control. By the data quality problem, some errors occur and it is increased at engineering sight.

Then, we found out several potential needs of improvement about BOM, PDM, and data quality.

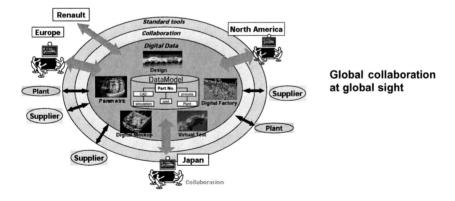

Global collaboration at global sight

Figure 3: Global collaboration at global sight

4.2 Requirements to the BOM system

A BOM system is the best index tool to manage parts list and its information through all vehicle life. Now, the Nissan BOM system manages only physical engineering release information but we need to expand the digital phase management by BOM system. We already started to use 3D CAD data in many divisions. Therefore we need a data management through all the vehicle life.

4.3 Requirements to the PDM data model

PDM is the database for digital validation that is required precise data management related with BOM system. PDM also needs to be accessed anywhere in the world development sight. Now Nissan uses the same PDM system in 3 major development sights in the world. It has got separated databases but they are synchronized between each other. We expect a single database to be the solution of the global concurrent development. The third requirement for PDM is direct access from suppliers. Now, many of the suppliers' engineers are working internal the Nissan R&D office. And the many of suppliers' colleagues is also working stay at suppliers' office as back office. For building an effective working environment, we should create a direct collaboration environment.

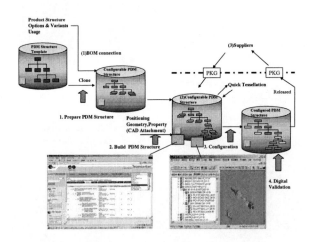

Figure 4: Requirement to PDM data model

4.4 Make the best use of viewing data

According to the progress of digital process, utilization of viewing data is required more and more. 3D CAD data became precise, and data size is increasing. Expanding of data size is impacted to response of CAD system operation. The other side, viewing data utilization is increasing. For example, styling visibility study, perceived quality study, DMDR study, manufacturing check in plant, service parts catalogue building, maintenance manual creation, and sales catalogue creation is used by viewing data.

4.5 Data quality standardization

The last issue is 3D CAD data quality. Compared with wire frame CAD models the data errors increase by using solid and surface models. One of the problems is created by engineers themselves. Engineers sometimes define contradiction shapes by themselves without any sense. In another case that problem occurs when the data are translated from other CAD system. Nissan uses the I-DEAS system but some of the suppliers are using other CAD system. They develop their parts by their own CAD system and finally they translate the data to I-DEAS format. It is one of the typical sources of error.

Nissan had built the data check rules and deployed. Nissan made a quality gate of 3D data and decided to accept just certificated data. By that the data quality should increase but it also one of the additional processes within our vehicle development. Now, using the same CAD system is more of a key subject when we develop the vehicle concurrently.

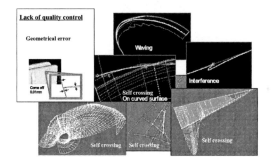

Data quality assurance

Figure 5: Data quality assurance

5 Conclusions

We believe that we have achieved a 3D data centric process with paperless drawing. We could reduce design change, development period, and development cost. The electronic release system also reduces the delivery time of engineering releases.
In the next 10 years, we have to build a global digitalization process for worldwide collaboration with alliance OEM and suppliers. Development anywhere, anytime, this is next target of IS infrastructure.

6 Author

Daisuke Shimomura, GLOBAL IS DIVISION SYSTEM DEVELOPMENT DEPARTMENT, Nissan Motor Co., Ltd., Atsugi-city Kanagawa, Japan
daisuke.shimomura@renault.com

PLM in Automotive Engineering
- today's capabilities & tomorrow's challenges -

Rainer Stark

Ford Werke AG, Cologne (Germany)

Abstract: This paper provides a conceptual overview of the current PLM landscape at Ford Motor Company (FMC) and an outlook of targeted changes to the Product Creation Solution. FMC has decided to substantially modify its virtual product creation solution. This includes taking existing solutions of the individual brands (Ford, Jaguar, Land Rover, Volvo and Mazda) to next level of modern virtual prototyping and developing disparate but into globally agreed virtual processes and practices. Within this context, the product creation solution comprises business and engineering processes, engineering tasks and virtual methods, as well as the underlying tools, applications and IT infrastructure. The paper concludes with general observations, which might apply to many OEMs and suppliers, as they continue in their concerted efforts to apply CAx and PLM solutions to replace, step-by step, physical prototyping by virtual prototyping.

1   FMC: Entrepreneurial and manufacturing pioneering

Ford Motor Company (FMC) entered the business world on 16th June 1903, when Henry Ford and his business associates signed the company's articles of incorporation. With $28,000 in cash, the pioneering industrialists gave birth to what was to become one of the world's largest corporations. Today, at the beginning of its second century of existence, FMC is an international enterprise with a worldwide organization and a true global customer base. It is the second biggest Automotive Producer worldwide, with approx. 340.000 employees, a business and customer base in more than 200 countries and approx. 114 manufacturing locations in 40 countries. FMC is also the biggest Automotive Finance Service Provider worldwide.

The global expansion of FMC has begun in the 60's. In 1967 the Ford of Europe was established and in 1971 the company established its North American Automotive Operations, consolidating U.S., Canadian and Mexican operations. With approx 38.000 employees, Ford-Werke GmbH (Germany) plays a leading role in the Ford of Europe organisation in hosting the European Headquarter in Cologne (Germany) and the following other facilities:

- Product Development Centre in Cologne
- Manufacturing and Assembly Plants in Cologne, Saarlouis (Germany) and Genk (Belgium)
- Ford Research Centre in Aachen (Germany)
- Test and Proving Ground in Lommel (Belgium)

FMC started with a single man envisioning products that would meet the needs of people in a world on the verge of high-gear industrialization. Today, FMC is a family of automotive brands consisting of: Ford, Lincoln, Mercury, Mazda, Jaguar, Land Rover, Aston Martin and Volvo.

Perhaps FMC's single greatest contribution to automotive manufacturing was the moving assembly line. It revolutionized mass production, lowering prices, helping the company far surpass the production levels of their competitors and making the vehicles more affordable. Today, there is another revolution underway – the Virtual Product Creation and Product Lifecycle Management (PLM), which is revolutionizing not only the assembly line but also engineering, design and prototyping. It's giving FMC the ability to shorten product development time and eliminate redundant testing and prototyping. It makes concurrent engineering possible, as well as better retention, dissemination and sharing of corporate knowledge.

From that very first assembly line in 1908 to the virtual technology of today, FMC has always been a leader in technology and innovation.

2 Technology transition for Virtual Product Creation

The technology transition for Virtual Product Creation in Automotive Business is the history of computer aided design and graphical computing. In the beginning of the 60's, i.e. the same time, in that Ivan Sutherland's Sketchpad revolutionized the world, the first computer-aided design program PDGS (Product Design Graphics System) entered the Automotive Business at FMC. Though, it was a long way from "imitating" a drawing table by producing 2D drawings at computers to advanced Solid Modelling, Product Data Management (PDM), Digital Mock-up (DMU), Virtual Reality (VR) and PLM. Today, nearly each process in the Automotive Business is being supported by an adequate software system (Fig. 1).

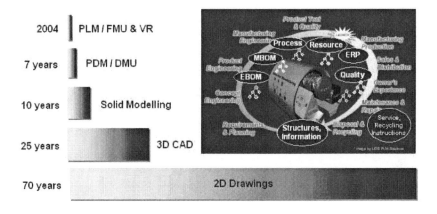

Figure 1: Technology transition for Virtual Product Creation

When looking at today's process of Virtual Product Creation, we consider the central role of CAx functionality (the software systems implementing it) and an even more important role of the corresponding Product Information Management mechanisms and environments. Since the individual software systems are able to cover specific tasks, the usage of different product information sets is a key to guarantee the realization of effective workflows and business processes.

Currently, the PLM landscape at FMC is under major "tune-up" for next levels, particularly with the challenge to migrate from existing PDM solutions towards the new integrated solutions.

3 The overall characteristics of PLM at FMC

As Ford Motor Company is a multi-brand corporation different legacy approaches to PLM exist. The mainly used approach (used by four brands) contains:

- Distinction between workgroup and enterprise PDM
- Distributed and fragmented Enterprise PDM for CAD & CAE data
- No integration between PDM, Business BOM (Bill Of Materials) and ERP (Enterprise Resource Planning) world
- Fully scalable supplier integration solution: file-based or asynchronous direct
- PDM triggered global Digital Mock-up generation and distribution
- Simple and static CAD product structure

On the other hand, there are further approaches used be one brand only. These contains:

- Distinction between workgroup and enterprise PDM
- Deep integration of Manufacturing virtual tool sets into PDM
- PDM-based Digital Mock-up generation
- No integration between PDM and Business BOM / ERP world

or

- Workgroup PDM only, fully integrated / driven by Business BOM / ERP
- PDM integrated Design in Context, local DMU caching only, configured Product Structure (PS)

or

- like above but without integration to Business BOM and ERP and with partly configured PS.

Multi-brand or partner development projects are seriously impacted by the diverse PLM approaches. The conversion between the individual approaches is difficult, costly and tedious. For example, Ford as a brand uses a specific legacy PLM solution called: *Separating and linking "Business PLM" and "Virtual PLM"* (Fig. 2), which has limitations in terms of integrated BOM and Virtual Product Configuration.

PLM in Automotive Engineering - today's capabilities & tomorrow's challenges -

In the Ford brand legacy PLM solution, the PIM (Product Information Management), which is the Ford's terminology for PDM, contains the Master Package Inventory (consisting of single, central managed master, including the latest Design Intent) implemented as a Metaphase structure.

Desired interfaces enable the bi-directional communication with several Team Data Management (TDM) systems. Unidirectional interfaces provide the Virtual Prototype and other visual information (for example for VisMock-up on PCs) to the users.

A mainly unidirectional interface allows data exchange with CAE applications. Obviously, the results of CAE simulations have to be brought back to the PIM structure.

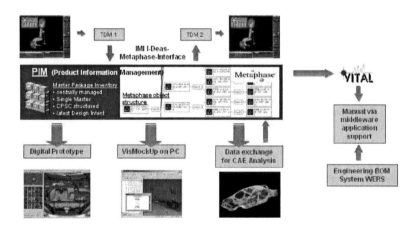

Figure 2: The concept of separating and linking "Business PLM" and "Virtual PLM"

4 The mega project C3P Next Generation

FMC is implementing PDM functionality through the C3P NG initiative. C3P stands for:

        Computer-Aided Design

        Computer-Aided Engineering

        Computer-Aided Manufacturing

        Product Information Management

The original C3P project put emphasis on PIM – the global access to current engineering and manufacturing information. It is also a platform of a distributed, component-based tool-set to support the integration of Computer-Aided Design and Manufacturing (CAD/CAM), Computer-Aided Engineering (CAE) and Knowledge Based Engineering (KBE) applications. These applications allow engineers worldwide to collaborate on the design process and fully optimise the design process.

The key Business Processes challenges to improve C3P are:
- PDM System and Design Process alignment through all the brands worldwide
- Change Management
- xBOM Management, allowing consistent EBOM (Engineering Bill of Materials, a list of the parts forming a product as determined by Design Engineering) and MBOM (Manufacturing Bill of Materials, a list of the parts, materials, and tools required in the manufacture of a product)
- Configured Digital Build, allowing digital Mock-ups that can be subjected to a wide variety of digital simulations, validations, tests and measurements.

Therefore, FMC established the Next Generation of the C3P initiative (C3P NG) to leverage the C3P standards and technology and integrate its suppliers into this environment. In addition to the above outlined key capabilities, C3P NG is targeted to provide the following additional extensions:
- Multi-CAD Design Management – the ability to manage and exchange design information authored under multiple CAD systems, including UG NX, I-DEAS, CATIA V4 and CATIA V5.
- Visual Collaboration – utilizing industry-standard JT Open technology to integrate data from dissimilar CAD systems into high-level assemblies and vehicle variants that can viewed, interrogated, marked-up and exchanged by widely dispersed users.

To manage engineering processes and all related product data exchange on an enterprise basis the C3P NG uses the capabilities of the Teamcenter PDM tool suite.

At the OEM's side, the digital data repository (e.g. CAD or Visualization Data) allows the downstream processes like Manufacturing and Services (for example to support

After Sales operations) the acquisition of the valid product data on-demand. Lightweight Visualization Data, e.g., allow to communicate complex product ideas to not yet sourced suppliers (avoiding exposure of sensitive details and intelligence). Already sourced suppliers can use the translation services to grab and deposit full-detailed CAD Data and/or Visualization Data with NURBS and PMI for the purposes of Design in Context.

5 The true PLM challenges in Automotive Engineering

In today's PDM systems, the product structure exists in two different occurrences (or representations) simultaneously:

- Encapsulated in CAD files, in the form of parametric models and, more or less, associative structures of CAD assemblies
- Exposed as a structure-tree (EBOM), with structure nodes which hold the possible configurations of the product

Since the encapsulated assembly structures are also a part of the product configuration, the permanent transition between these two representations becomes crucial.

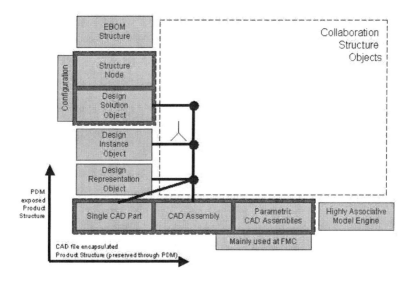

Figure 3: Exposed vs. encapsulated Product Structure in a PDM system

To keep the virtual EBOM consistent is of extreme importance for the penetration of the product structure into downstream and business processes. Along that line another true challenge of PLM exists: How to keep the virtual EBOM, the virtual MBOM and the Engineering Collaboration Structure consistent as a basis for integrated ERP, Business and Logistics processes? The Product Data in form of CAD models, as well as Plant Resources (Tools) and their CAD models build in turn the foundation for the xBOM and Engineering Collaboration structures. Collaboration Structures (as indicated in Figures 2 and 3) are needed to provide a flexible approach to manage product descriptions which need to be handled differently to the structure of the BOMs, e.g. to represent associative models for specific CAD authoring or for pure ad-hoc proposal sharing between organization (i.e. not yet clear whether they will be linked into BOM representations of the vehicle).

At the implementation of the Virtual Product Creation and PLM as demanded by the challenges mentioned above, very different core competency and virtual development skills are needed. At this point, the world of the Automotive Engineering meets Software Engineering.

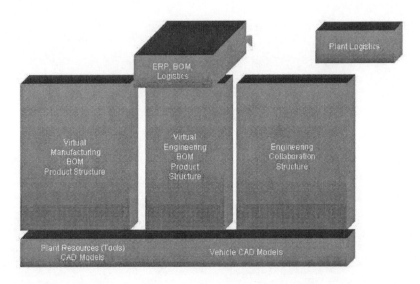

Figure 4: The true PLM challenge in the Automotive Engineering

The established world of Automotive Engineering, with more than 100 years of history and well-defined job functions ranging from Craftsman to Chief Engineers is driven by appearance and the attributes of the product (i.e. a car). The homologation and public judgment of the product takes place before releasing it to the customer. The delivery to the customer requires all components and systems of a product to be finished and ready for operation.

In opposite, the young discipline of Software Engineering, with approx. 30 years of history, is still partially shaped by gurus and "artists". The job functions are ranging from specialists to professional Project Managers, the development of a product (a software system) is algorithm and modularisation driven. But perhaps the biggest problem is, that the product is delivered usually with 2-3 test cycles at the key-users (i.e. customers!) and it is almost standard, that not everything is available at the product delivery time (workaround mentality).

Anywhere in between of these two worlds emerges continuously the "brave" world of Virtual Product Creation, which is difficult for traditional Automotive Management to understand and to assess. It is multi-disciplinary, affects processes, methods and tools and occurs in turbulent development cycles.

The following "delivery approaches" picture the changing relationship between the core business and the software development in the Automotive Engineering through the last decades:

- 60's and 70's: software application by automotive OEMs ("in-house" solutions)
- 80's: automotive OEMs to incorporate turn-key systems by CAD/CAM vendors
- 90's: outsource "entire" solutions to main vendors (with 3rd party tools)
- 2000+: technology supports multi-vendor solutions, the role of solution integration becomes crucial

A general trend can be seen in this evolution: The Automotive Engineering is moving away from the focus on pure functionality or IT systems towards seamless enterprise solution integration.

The side-effect of the changing approaches to the implementation of software solutions in the Automotive Business is, that the number of experienced experts in Product Development and Manufacturing, as well as Service Engineering is very small at

OEMs but relatively high outside the core automotive engineering in the role of Application or Consulting Experts.

The outsourcing of "entire" solutions to the main vendors and 3rd party developers has led to the situation, that PLM consulting can be offered by many partners, but scalable PLM solutions can be provided by a few partners only and are not available off-the-shelf.

6 Conclusions

- Virtual Product Creation and PLM becomes core backbone for efficient development process and reduced time to market.
- PLM landscape at Ford Motor Company (and its competitors) is under major "tune-up" for next levels. The challenge is to migrate from existing PDM solutions to enterprise-wide integrated solutions.
- Multi-brand and partner development projects at FMC are seriously impacted by diverse PLM approaches. The conversion between the approaches is difficult, costly and tedious.
- OEMs to prepare major paradigm, core competency and virtual development skill-set change in the work force.
- Many partners offer PLM consulting, but scalable PLM solutions can only be provided by a few partners and are not available off-the-shelf.
- The group of experienced experts in Product Development and Manufacturing, as well as Service Engineering is very small at OEMs – the number of experts outside the core automotive engineering areas is relatively high.
- Solution integration and penetration into Engineering and Business is the key challenge.

7 Author

Dr. Rainer Stark, Ford Werke AG, Spessartstrasse, 50725 Cologne, Germany, rstark1@ford.com

Digital Colour Process for Automobile Design

Atsushi Takagi [1]; Shimada Hiroaki [2]

1) Toyota Motor Corporation, 1, Toyota-cho, Toyota-city, Aichi, 471-8571 JAPAN
2) Nihon Unisys Solutions, Ltd., 1-1-1 Toyosu, Koto-ku, Tokyo, 135-8560 JAPAN

Abstract: We have developed a Computer Graphics (CG) system, DSR, to satisfy designers' requirements in automobile development. DSR joined know-how all-out effort to support styling and colour design by analyzing automotive design process. DSR is able to reproduce accurate colour and material. We introduce application example in some stages of automobile development process and technologies have used.

1 Introduction

With the advancement of shape modelling techniques to deal with free-form surfaces, CAD was in practical use for vehicle style design since the beginning of the 1980's[1]. In this type of CAD, the mathematical model constructed inside the computer is usually converted into a clay model by a NC-milling machine for three-dimensional evaluation. Meanwhile, application of the CG rendering technique has remarkably advanced in recent years making it possible to evaluate shapes and colours on a graphics display before making a physical clay model.

However, because colours and shapes are more strictly evaluated by designers, the requirements for practical use of CG for car design are very severe.

This paper proposes a CG rendering and interactive technique that meets designers' requirements. To satisfy the requirements, we have developed DSR.

2 DSR

DSR stands for "Digital Styling Review". This is the Computer Graphics (CG) System joined know-how all-out effort to support styling and colour design by analyzing automotive design process. DSR is able to reproduce accurate colour and material.

2.1 Features of DSR

DSR has two features. One is its quality (

Fig. 1). DSR is a simulation system for colour and styling. Output image has accurate colour[8], [9], [10] and accurate material. Because of using measured reflectance, we do not need to adjust colour and material manually. And colour should be accurate in order to utilize design field. For example, colour on TV and colour of paper should be equal, in order to evaluate automobile exterior colour. On the other hand, commercial CG systems need to express colour and material by hand. Accordingly, colour on CRT is generally not accurate like

Fig. 1.

Second, The difference is Shape format in system(

Fig. 2). DSR can handle free form surface. That is, DSR can execute Ray Tracing on free form surface directly. On the other hand commercial CG system can not teat free form surface directly. Free form surface should be transferred to polygon surface. For this reason, The direction of reflected light or refracted light is not accurate. In this case, we could not find installation hole because of inaccuracy of ray tracing.

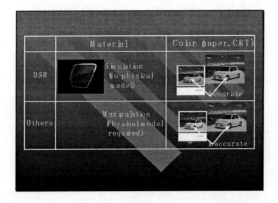

Fig. 1 Quality

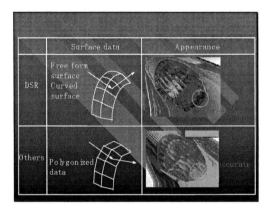

Fig. 2 Surface Data in System

2.2 System configuration of DSR

Fig. 3 is a System configuration of DSR[4]. DSR is a CG system which works on PC. Styling data is brought from another CAD system. DSR takes styling data as IGES or TOGO[1] native format. Configuration of animation, rendering, motion are DSR's task. DSR utilizes measuring colour data, and external motion dynamics system such as ADAMS. Rendering is heavy, so it generally executed by external rendering server. As output images of DSR is very high quality, we use in-hause colour hard copier as figure shown. This colour hard copier is able to have 9 primary inks. Now we use 7 colour inks, Y, M, C, K, R, G, B. This can reproduce all automotive exterior colours[11], [12].

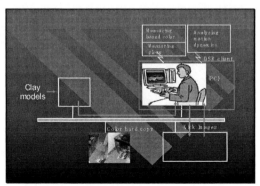

Fig. 3 System Configuration of DSR

3 Application Examples

3.1 Design Stage

Fig. 4 is a CG simulation of exterior evaluation. Car is produced by DSR. Background is real. We can evaluate styling and colour accurately like this image.

Fig. 4 CG Simulation of Exterior Evaluation

Fig. 5 is CG simulation of interior evaluation. Iinterior is more difficult as compared to exterior. We use several technologies for rendering such as global illumination model, fabric texture, such as photon mapping, etc.

Fig. 5 CG Simulation of Interior Evaluation

3.2 Production Engineering Stage

Next, we would like to show the examples in production engineering stage.

Fig. 6 is CG simulation of small parts.

Fig. 7 is the lamps and reflectors.

Fig. 8 is a simulation of optical thin film. Outer lens is coated by thin filter. We calculate colour of lens taking account of various phenomena caused by tunnel effect, resonance scattering, absorption and other effects.

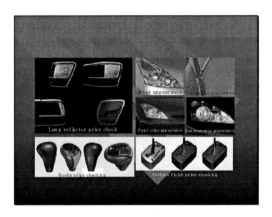

Fig. 6 CG Simulation of Small Parts

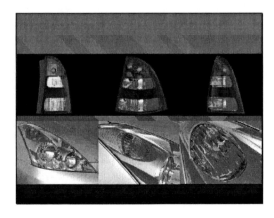

Fig. 7 CG Simulation of Lamps

Fig. 8 CG Simulation of Optical Thin Film

3.3 Advertisement & Marketing Stage

Next, we would like to show the examples in advertisement & marketing stage. In this stage, we are forced to use CG instead of actual car in the near future because of lack of prototype models which have been used for commercial films etc.

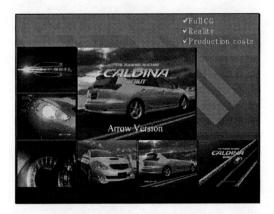

Fig. 9 CG Simulation of TV Commercial Film

Fig. 9 is a example of TV commercial film. This is the first version we released on TV. We released this on September of 2003.

Digital Colour Process for Automobile Design

4 Rendering Model[3]

In this section, the reflection model expression, which is the fundamental model for rendering, is first established. Next, the light source and material which are important factors for realism are modeled, and then they are integrated into the reflection model.

4.1 Reflection Model Expression

＜other work＞

The conventional expression of object surface reflection is as follows[2]

$$I(\lambda) = \varepsilon(\lambda) + \int_\Omega \rho(\lambda) L(\lambda) \cos\theta d\omega. \quad (1)$$

where,

- λ : Wavelength
- $I(\lambda)$: Reflected light going in a certain direction
- $\varepsilon(\lambda)$: Light emitting from an object going toward the direction of the reflection
- $\rho(\lambda)$: Reflectance of the object's surface
- $L(\lambda)$: Incident light
- θ : Incident angle
- $d\omega$: Differential solid angle of incident light
- Ω : Solid angle of the entire incident light

(See **Fig. 10**)

Though this is an excellent expression for a global illumination, it is not clear on the following points.

(P1) $\rho(\lambda)$ is not clearly defined. Because there are several different types of object surface reflectance, it is necessary to clarify which type to apply.

(P2) $L(\lambda)$ is not minutely analyzed.

(P3) Because the unit of each value is not clear, comparison with the actual value cannot be made. Therefore, it is difficult to verify the actual data.

<Our approach>

We have solved the problems described in the above Items (1) through (3) and propose the following expression for practical use.

$$I(\lambda) = \varepsilon(\lambda) + \frac{1}{\pi}\int_\Omega \beta(\lambda)L(\lambda)\cos\theta d\omega$$
$$(W \cdot sr^{-1} \cdot nm^{-1} \cdot m^{-2}) \qquad (2)$$

Where, λ is the wavelength having unit (nm). $I(\lambda)$, $\varepsilon(\lambda)$, and $L(\lambda)$ have the same definition as that of expression (1). Each of them is the spectral radiance having each unit $(W \cdot sr^{-1} \cdot nm^{-1} \cdot m^{-2})$. Ω, θ, and ω have the same definition. $\beta(\lambda)$ is the spectral radiance factor of object surface. Actually, however, the spectral reflectance factor $R(\lambda)$ having unit (1) realizing colourimetry is used. Where, $R(\lambda)$: spectral reflectance factor (1) from $L(\lambda)$ to $I(\lambda)$ The coefficient $\frac{1}{\pi}$ before the integral sign appears because of using the spectral radiance factor $\beta(\lambda)$ as the reflectance. This expression includes the following to solve the problems (P1) through (P3).

1] All units are clarified and the measured value $\beta(\lambda)$ or $R(\lambda)$ is used as $\rho(\lambda)$.
2] Especially the direct sunlight and sky light are precisely calculated for $L(\lambda)$.

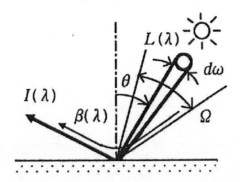

Fig. 10 Incident light $L(\lambda)$ and reflected light $I(\lambda)$

Digital Colour Process for Automobile Design

This paper advances by omitting $\varepsilon(\lambda)$ for convenience sake.
That is,

$$I(\lambda) = \frac{1}{\pi} \int_\Omega \beta(\lambda) L(\lambda) \cos\theta \, d\omega \quad \left(W \cdot sr^{-1} \cdot nm^{-1} \cdot m^{-2} \right) \quad (3)$$

To clarify the expression (3) separate the incident light $L(\lambda)$ into specular reflected light, direct sunlight, and other light sources.

$$\Omega = \Omega_R + \Omega_S + \Omega_G \quad (sr) \quad (4)$$

Where,

Ω_R : Solid angle of the area in the direction of specular reflection
Ω_S : Solid angle of the area in the direction of main light source
Ω_G : Solid angle of the area other than the above

The above expression is illustrated in **Fig. 11**.

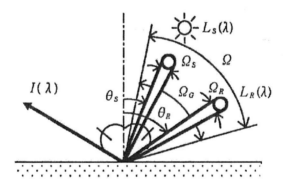

Fig. 11 Division of Ω

Where each value is defined as follows :

$L_R(\lambda)$: Spectral radiance of the light from the direction of specular reflection
$L_S(\lambda)$: Spectral radiance of main light source (direct sunlight)
θ_R : Incident angle (rad) of $L_R(\lambda)$
θ_S : Incident angle of $L_S(\lambda)$
$\beta_R(\lambda)$: Spectral radiance factor (1) from $L_R(\lambda)$ to $I(\lambda)$

$\beta_S(\lambda)$: Spectral radiance factor (1) from $L_S(\lambda)$ to $I(\lambda)$

$d\omega_R$: Differential solid angle (sr) of $L_R(\lambda)$

$\Delta\omega_S$: Solid angle of $L_S(\lambda)$

Therefore, the reflected light $I(\lambda)$ is obtained from the expression (3) as follows :

$$I(\lambda) = \frac{1}{\pi}\int_{\Omega R+\Omega S+\Omega G}\beta(\lambda)L(\lambda)\cos\theta d\omega$$

$$= f(\lambda)L_R(\lambda) + \frac{1}{\pi}\beta_S(\lambda)L_S(\lambda)\cos\theta_S \cdot \Delta\omega_S + G(\lambda).$$

(5)

However, assume $f(\lambda)$ and $G(\lambda)$ are as follows :

$$f(\lambda) \equiv \frac{I(\lambda)}{I_R(\lambda)} = \beta_R(\lambda)\cos\theta_R d\omega_R. \qquad (6)$$

$$G(\lambda) \equiv \frac{1}{\pi}\int_{\Omega G}\beta(\lambda)L(\lambda)\cos\theta d\omega. \qquad (7)$$

For normal paint surfaces, because $f(\lambda)$ can be approximated by the fresnel factor f which is independent of wavelength, the expression (5) can be rewritten as follows :

$$I(\lambda) = f \cdot L_R(\lambda) + \frac{1}{\pi}\beta_S(\lambda)\cos\theta_S \cdot \Delta\omega_S + G(\lambda). \qquad (8)$$

4.2 Light source

4.2.1 Direct sunlight

Because direct sunlight has been ambiguously considered so far, it has been used only as a parallel light source with highluminance. Habu et al. insist that direct sunlight depends on the latitude, season, hour, air contamination, and amount of steam at the arrival point and has the characteristic in which the variation depends on the wavelength. Therefore, because it is assumed that direct sunlight influences the appearance of an object, we decided to rigorously consider direct sunlight. The spectral

Digital Colour Process for Automobile Design

radiance $L_S(\lambda)$ of sunlight on the ground surface is expressed by the spectral irradiance $E_m(\lambda)(W \cdot nm^{-1} \cdot m^{-2})$ on the ground surface and its solid angle $\Delta\omega_S(sr)$ as the following expression.

$$L_S(\lambda) = E_m(\lambda)/\Delta\omega_S \qquad (W \cdot sr^{-1} \cdot nm^{-1} \cdot m^{-2}) \qquad (9)$$

Moreover, $E_m(\lambda)$ is expressed according to the Beer-Bouguer-Lambert's law as follows:

$$E_m(\lambda) = E_0(\lambda) \cdot e^{-(C_R(\lambda)+C_M(\lambda)+C_{OZ}(\lambda))m} \cdot \tau_0(\lambda) \cdot \tau_W(\lambda) \qquad (W \cdot nm^{-1} \cdot m^{-2}) \qquad (10)$$

Where,

$\Delta\omega_S$: Solid angle when viewing the sun from the ground surface

m : Air mass

$E_0(\lambda)$: Spectral irradiance out of atmosphere. In this case, use CIMO-VII(1981)

$C_R(\lambda)$: Attenuation factor according to Rayleigh diffusion of air molecules

$$C_R(\lambda) = 0.00864\lambda^{-(3.916+0.074\lambda+0.050/\lambda)} \qquad (\lambda : \mu m) \qquad (11)$$

$C_M(\lambda)$: Attenuation factor according to diffusion of aerosol

$$C_M(\lambda) = \beta\lambda^{-\alpha\lambda} \qquad (\lambda : \mu m) \qquad (12)$$

The factors α and β are obtained through measurement.

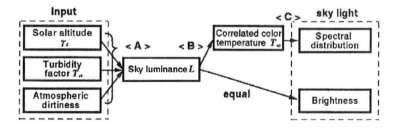

Fig. 12 Process to Determine Spectral Distribution of SkyLight

$C_{0z}(\lambda)$: Absorption coefficient due to absorption of ozone

$\tau_0(\lambda)$: Attenuation factor according to oxygen molecules in the atmosphere

$\tau_w(\lambda)$: Attenuation factor according to water vapor in the atmosphere

Among the above values, $\tau_0(\lambda)$ and $\tau_w(\lambda)$ are omitted because they are very weak in visible wavelength, and their influence is small.

4.2.2 Sky light

<Other work>

Our purpose for sky light integration is to clarify spectral distribution and luminance at each point in the sky under any weather condition. Many research studies have been made about luminance so far. For spectral distribution, however, no decisive research has been made. This is because the logical analysis of spectral distribution is difficult since sky light is generated through complex processes such as dissipation of direct sunlight due to air, aerosol, and clouds or repetitive reflection of it on the ground. The following are examples of relevant research. Nakamae et al.(1980) simulated object appearance with CG under clear sky and overcast sky using the CIE standard clear sky light luminance expression. R.V. Klassen(1987) simulated refraction, diffusion, and absorption due to air molecules and aerosol to express the colour of sky light through CG. Sekine(1989) analytically calculated spectral distribution at each point of clear sky In these research studies, however, spectral distribution a-teach point in the sky under any weather condition was not obtained.

<Our approach>

We proposed a method to obtain spectral distribution at any point in the sky under any weather condition using the empirical formula obtained through measurement.

[Item 1] Classification of sky light We clarified the sky that we handled. In general, the sky can be classified into the following three types.

<1> Clear sky: Sky free from clouds

<2> Intermediate homogeneous sky: Sky in which weather homogeneously changes between clear and overcast skies without clouds scattered in the sky

<3> Overcast sky: Sky covered with clouds so thick that the sun cannot be seen

Digital Colour Process for Automobile Design

In the intermediate sky expressing the intermediate state between clear and overcast skies, clouds are actually scattered. In this section, however, we use the intermediate homogeneous sky in Item <2> for convenience sake. Also in this section. we deal with Items <1> through <3> in order to meet.

[Item 2] Calculation of spectral distribution at any point of sky light. The method proposed by us uses the fact that sky light luminance has a certain correlation with colour temperature(See **Fig. 14**). The spectral distribution at any point of sky light is obtained as shown in **Fig. 12**. The process to determine spectral distribution of sky light, turbidity factor T~,, and atmospheric dirtiness are described in 4.2.3. The calculation for processes <A>, <B>, and <C> in **Fig. 12** is defined as follows :

Process <A> (Calculation of sky light luminance distribution)
The luminance distribution at each point in the sky can be obtained by inputting the sun altitude γ_s, turbidity factor T_{vl}, and atmospheric dirtiness. The following luminance expressions are proposed corresponding to the skies<1>, <2>, and <3> classified in [Item 1].

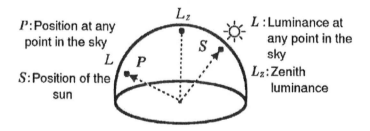

Fig. 13 Symbols in the Sky

<1> CIE standard clear sky light luminance function
<2> Intermediate homogeneous sky light luminance function
<3> CIE standard overcast sky light luminance function

The above expressions in Items <1> and <3> are specified as international standards, which express only clear sky and overcast sky respectively. The luminance at each point of the sky is expressed as the ratio to the zenith luminance $L_z (cd/m^2)$.

Though the expression in Item <2> is not recognized as an International standard yet, it is shown in the 1988 CIE Technical Report in detail. It is possible to continuously change the luminance at each point of the sky between clear and overcast using the turbidity factor T_{vl} as a parameter. Items <1> through <3> are expressed in the following form respectively.

<1> $L/L_z = f_1(P,S)$ (1)   (13)

$L_z = g_1(T_{vl}, \gamma_s)$ (cd/m^2) (14)

<2> $L = f_2(P, S, T_{vl})$ (cd/m^2) (15)

<3> $L/L_z = f_3(P,S)$ (1)   (16)

$L_z = g_2(\gamma_s)$ (cd/m^2) (17)

Process <B> (Conversion of luminance into correlated colour temperature)
When assuming the luminance at any point in the sky as $L(cd/m^2)$ and the correlated colour temperature corresponding to the above luminance as $T_{cp}(K)$, the relationship between L and Tcp is shown by the expression below,

$$T_{cp} = \frac{1.1985 \times 10^8}{L} + 6500 \qquad (K). \qquad (18)$$

The above expression is based on our measurements. That is, we regressively obtained it from the results of examining the relationship between the luminance at any point in the sky and colour temperature at any hour and spot in Japan. We found that there is a strong link between them(See **Fig. 14**).

Process <C> (Conversion of correlated colour temperature into spectral distribution) A method is needed to convert correlated colour temperature into spectral distribution using measured data. However, because the method is not determined at present, we used the CIE synthesized daylight expression proposed by Judd, MacAdam, Wyszecki

4.2.3 Weather simulation

A car body's surface colour is evaluated by assuming its appearance at various places and under various weather conditions, because the surface may not look good under a cloudy sky though it looks good under a clear sky. For the existing CG, most research is made for the rendering of appearance under a clear sky with direct sunlight, and does not include the idea of a rigorously rendered appearance depending on weather condition. We proposed a method to render the appearance of objects under any weather condition. For this, we first considered the spectral distribution and intensity of direct sunlight and skylight which are the main outdoor light sources by defining the factors which determine weather and using the factors as input data.

[Item 1] Factors to simulate weather It is considered that the following four factors change weather. (1) Position (Latitude and longitude on the earth). (2) Date and hour (Hour based on universal time). (3) Weather simulation factors. They are the values defined by meteorology. <1> Atmospheric transmittance: P, P_v. <2> Coefficient of turbidity α, β, water vapor content, oxygen content, and ozone content. <3> Turbidity factor: T, T_{vl}. (4) Atmospheric dirtiness. This is the parameter to indicate how the atmosphere is contaminated due to artificial factors, though this is not minutely studied at present. For sky light, however, there is the method to indicate the condition of the sky on contaminated industrial areas by correcting the sky light luminance distribution.

[Item 2] Calculation for direct sunlight and sky light.
  (1) Direct sunlight. For direct sunlight, $E_m(\lambda)$ in the expression (10) can be obtained using the factors described in [Item 1] to obtain the direct sunlight value $L_S(\lambda)$ as a light source using the expression (9).
  (2) Sky light. For sky light, it is first necessary to know the Luminance distribution of any weather condition. In this case, we use the intermediate homogeneous sky light Luminance function in the expression (15). This can also be obtained using the factors described in [Item 2]. P, P_{vl}, T, and T_{vl} can be obtained with $E_m(\lambda)$ and air mass m

4.3 Material

To express realism, a rigorous definition of materials as well as light sources are needed. Because the modality of reflection or transmission when light hits an object depends on the object, it is very important for expressing appearance of various materials to accurately check these physical phenomena. Reflection or transmission produced when light hits an object is generally classified into the following four types from ① through ④.

①Diffuse reflection

Most existing research mathematically models the diffusion reflection according to Lambert's law. However, it is difficult to use these model expressions because they include several parameters to delicately change the material, and the parameters

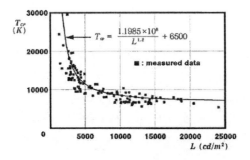

Fig. 14 Measured Data and Approximate Formula

must be adjusted through trial and error in order to express a proposed material. Therefore, the model expressions are rarely applied to the requirement to accurately render the materials of existing objects. Prof. Minato has established the method to specify the diffuse reflection of objects by measuring the spectral reflectance factor showing the diffuse reflection. By using the reflectance obtained through measurement as the reflectance $\beta(\lambda)$ in the expression (3), the material of an actual object can accurately be rendered without setting parameters. The spectral reflectance factor $R(\lambda)$ is measured by changing the angle (aspecular-angle) on the basis of the specular reflection direction as shown in **Fig. 16**. **Fig. 17** shows the equipment sused to measure $R(\alpha,\lambda)$. The results of measuring $R(\alpha,\lambda)$ for paint are shown in **Fig. 19**. The material can be decided according to the shape of the graph.

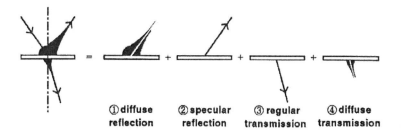

① diffuse reflection   ② specular reflection   ③ regular transmission   ④ diffuse transmission

Fig. 15 Classification of Reflection and Transmission

Therefore, various materials can be created by controlling and changing the graphic shape. We further developed the method of Minato et al. in order to accurately perform the colouring calculations for any weather condition. That is, we devised the method to calculate $G(\lambda)$ in the expression (7) depending on the material. $G(\lambda)$ is an important term to determine the brightness depending on background light. For example, because $R(\alpha,\lambda)$ of a solid paint surface can be considered as a homogeneous diffusion surface taking an almost constant value when α goes beyond the ranges of 10< aspecular angle <20 deg., it is possible to integrate only ω by removing $R(\alpha,\lambda)$ from the integral terms. Meanwhile, because the change rate of $R(\alpha,\lambda)$ to α is large for a metallic paint surface, it is necessary to accurately calculate $G(\lambda)$ for ω. Especially under a cloudy sky light without direct sunlight, it is not permitted to simply approximate $G(\lambda)$ because the weight of $G(\lambda)$ is large in the colouring calculation expression (5).

Moreover, because the area of small a (about 0< aspecular angle <30 deg.) where $R(\alpha,\lambda)$ steeply changes and greatly influences the appearance of objects under a cloudy sky, it is necessary to accurately measure the area.

② Specular reflection

Specular reflection is the factor to determine the reflecting degree of lustrous objects, which is an important factor contributing to realism. Therefore, it is necessary to strictly consider the specular reflection as well as the diffuse reflection. The specular reflection is expressed as the specular reflectance. Because the specular reflection depends on the material, it should be determined for each material. The specular reflection of a pure glass surface producing no diffuse reflection follows Fresnel's formula. However, it is difficult to formulate the specular reflection because the materials, some paint surfaces, metal surfaces, resin surfaces, and lustrous rubber surfaces considered in this section show characteristics different from Fresnel's formula. Therefore, it is most effective to obtain the specular reflectance through measurement.

For specular reflectance, the influence of light polarization is easily overlooked. However, specular reflectance also depends on the polarized-light condition of incident light. Therefore, to be exact, it is necessary to consider the specular reflectance including the influence of polarized light. For example, blue sky light in clear sky is clearly polarized. Therefore, when considering the reflection of the above light, the above concept is necessary. We define the extended Fresnel's formula capable of considering the polarization characteristics of incident light.

That is,

$$f = \frac{pr_s^2 + sr_p^2}{s+p}. \tag{19}$$

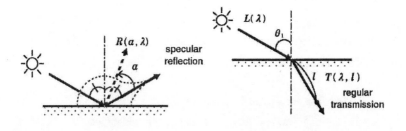

Fig. 16 $R(\alpha, \lambda)$ Fig. 17 $T(\lambda, l)$

Digital Colour Process for Automobile Design

Fig. 1Fig. 18 Gonio-spectro Photometer

Where, γ_s, and γ_s. represent the amplitude reflectance of Fresnel's s-wave and p-wave respectively, and s and p represent the energy ratio between s-wave and p-wave components of incident light respectively. Only when incident light is natural light, s equals p and the expression (19) is expressed as the following formula which has widely been used.

$$f = \frac{1}{2}\left(\gamma_s^2 + \gamma_p^2\right) \tag{20}$$

③ Regular transmission

For objects to transmit light like glass, it is necessary to obtain regular transmittance through measurement. When assuming regular transmittance as, t as is generally known, t can be expressed as $t = 1 - f$ using the specular reflectance f. In addition, because the transmitted light is attenuated inside the glass, the colour seen through the glass depends on the glass. Therefore, attenuation of transmitted light should also be considered. If light $L(\lambda)$ $(W \cdot sr^{-1} \cdot nm^{-1} \cdot m^{-2})$ enters chromatic or achromatic glass with the incident angle of θ_1, the transmitted light $T(\lambda, l)$ when the light $L(\lambda)$ advances by the distance $l(m)$ through the glass is expressed as following :

$$T(\lambda, l) = L(\lambda) \cdot t \cdot e^{-c(\lambda)l} \quad (W \cdot sr^{-1} \cdot nm^{-1} \cdot m^{-2}) \tag{21}$$

Where, $C(\lambda)$ represents the absorption coefficient of glass. $C(\lambda)$ can be obtained through measurement with the equipment shown in **Fig. 18**.

④ **Diffuse transmission**

Though the glass to diffuse-transmit light is rarely used for cars, it is widely used in the world, in which frosted glass is a typical example. To express a material like this, it is necessary to strictly consider diffuse transmission

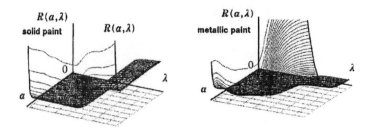

Fig. 19 Configuration of $R(\alpha,\lambda)$

5 Technologies

5.1 Background Composition [5], [6], [7]

Next, we will talk about background synthesizing technique. We need to synthesize real background images and CG cars. This means perspective of CG car should be equal to that of background. We can get perspective information, that is, eye point, object point and angle of field from these sequential image cells. We calculate projection matrices, and synthesize real background images and CG cars.

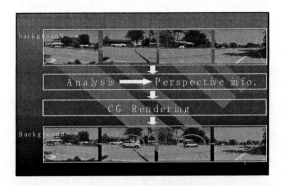

Fig. 20 Background Coposition

5.2 Motion Dynamics

Next, we will talk about background motion dynamics. We use a ADAMS CAR for motion analysis. This is a commercial motion analysis system. But we are not aware of ADAMS. We can handle motion analysis by using DSR commands. So we can control ADAMS on DSR easily.

6 Summary

We would like to summerive as follows:
1. DSR is a CG System which
   - has accurate colour and material based on measurement,
   - is based on ray tracing with free form surface.
2. DSR is heavily used,
   - in design, production engineering, advertisement stage,
   - to eliminate physical model or supplement of physical model.

7 Future Works

There are still some future works remaining.
1.Providing DSR to industrial design market.
2.Make DSR as a virtual prototyping tool for all design process.
3.Standardization of 5 angles method
4.Utilize DSR in virtual showroom application.

8 References

[1] Higashi Masatake, Kohzen, Ikujiro, Nagasaka, Junji, An Interactive CAD System for Construction of Shape with High-Quality Surface, E. A. Warmann(ed.), Computer Applications in Production and Engineering (North Holland), pp.371-390(1983)

[2] Kajiya, James Thomas, The Rendering Equation, Computer Graphics (Proceeding of ACM SIGGRAPHVol. 20, No. 4, August, pp. 143-150 (1986)

[3] Atsushi Takagi, Hitoshi Takaoka, Tetsuya Oshima and Yoshinori Ogata, Accurate Rendering Technique Based on Colourimetric Conception, Computer Graphics(Proceedings SIGGRAPH'90), Vol. 24, No.24, pp.263-272 (1990)

[4] Oshima, Tetsuya, Yuasa, Shinji, Sakanoshita, Ken-ichi, Ogata, Yoshinori, A CAD System for Car Design, Eurographics'92,Vol. 11, No. 3, `C-381'--`C-390' (1992)

[5] Gideon P. Stein and Amnon Shashua, Member, IEEE, Model-Based Brightness Constraints:On Direct Estimation of Structure and Motion, IEEE TRANDSCTIONS ON PATTERN ANALYSIS AND MACHINE INTELLIGENCE, Vol 22, No. 9, September 2000, pp. 992-1015 (2000)

[6] Olivier Faugeras, Three-Dimentional Computer Vision , A Geometric Viewpoint --, The MIT Press, Tird printing, (1999)

[7] E. Nakamae, X. Qin, G. Jiao, P. Rokita and K. Tadamura: Computer-generated still images composited with panned/zoomed landscape video sequences, Visual Computer, Vol. 15, pp. 429-442 (1999).

[8] Takaoka, Hitoshi et al, New Input Modulation Method for Generating Expected Colours on a CRT Monitor, SID '91, pp.57-60 (1991)

[9] Atsushi Takaghi, Toru Ozeki, Tetsuya Oshima and Yoshinori Ogata and Sachie Minato, An Inverse Problem Solutionwith Neural Network System, IPES-94 2nd International Symposium, pp. B44-B45 (1994)

[10] Atsushi Takaghi, Toru Ozeki, Hitoshi Takaoka, Tetsuya Oshima and Yoshinori Ogata and Sachie Minato, A Colour Reproduction Method with Neural Network Sytem, IS & T Tenth InternationalCongress on Advances in Non-Impact Printing Technologies, pp. 584-586, October 30--November 4, New Orleans (1994)

[11] Atsushi Takaghi, Toru Ozeki, Yoshinori Ogata and Sachie Minato, Faithful Colour Printing for ComputerGenerated Image Syntheses with Highly Saturated ComponentInks, The Second IS & T/SID Colour Imaging Conference --Colour Science, System and Applications--, pp.108-110, November 15--November 18, Scottsdale (1994)

[12] Atsushi Takaghi, Toru Ozeki, Tetsuya Oshima and Yoshinori Ogata and Sachie Minato, A Colour Printing System Enabling Reproduction ofthe Desired Colour: Widend Colour Gamut Realized by a Multiple Ink Method with the Aid of an Inverse Problem Solution, Computer Graphics --Development In Virtual Environment--, Academic Press, pp.131-157 (1995)

9 Authors

Atsushi Takagi, Toyota Motor Corporation, 1, Toyota-cho, Toyota-city, Aichi, 471-8571 JAPAN, takaghi@mail.toyota.co.jp

Hiroaki Shimada, Nihon Unisys Solutions, Ltd., 1-1-1 Toyosu, Koto-ku, Tokyo, 135-8560 JAPAN, Hiro.Shimada@usol.co.jp

Part III: *Product Definition*

Status of the ISO TC184/SC4 Parametrics Standards Development

Akihiko Ohtaka

Nihon Unisys, Koto-ku Tokyo, Japan

Abstract: Exchange or share of product information is critically important for collaborative product development, but today's level of data exchange is limited to snapshot explicit representation of product data. For fulfilling strong industry requirements to share design intent or design history, the ISO TC184/SC4 Parametrics group has developed related standards most of which are close to completion. This paper reports major contents and development status of standards under development.

1 Introduction

The motivation for parametrics work stems from developments in the history of CAD. Around 1980, there were two different state-of-the-art flavours of 3D shape modelling, one boundary representation and the other procedural representation. The former concentrated on the precise geometric details of the shape, and provided means for building a representation of that shape in an incremental manner. The operations provided were often concerned with building new elements into the shape model.

The latter concentrated more on high-level operations used to build the shape, and only generated an explicit geometric model when needed. Constructive Solid Geometry (CSG) was an early example of such a method, and a rigorous underlying theory was developed for it. In 1984, STEP development started. At that time pure CSG was on the wane, and general procedural representation was on the rise. Boundary representation was state of the art. Today, all the modern CAD systems use a dual procedural/B-rep approach.

A. Ohtaka

B-rep and procedural models have different complementary characteristics.

<B-rep>

1) Provides explicit geometry for applications (e.g., visualization, NC machining)
2) Relative positioning is easy
3) Parameterization is hard
4) Attribute association is easy
5) Representation is verbose
6) Maintenance of representation is delicate <Procedural>
1) Provides just operations. No geometry included which is no good for downstream applications
2) Relative positioning is hard
3) Parameterization is easy
4) Attribute association is hard
5) Representation is concise
6) Maintenance of representation is robust

Nowadays, it is quick and easy to generate a B-rep from a procedural model, and therefore we can get the best of both worlds.

Figure 1 illustrates the difference of these two representations.

1. c4 e6
2. Nc3 Bb4
3. Nf3 0-0
4. Bg5 c5
5. e3 cxd4
6. d4 Nf6
7. exd4 h6
8. Bh4 d5
9. Rc1 dxc4
10. Bxc4 Nc6
11. 0-0 Be7
12. Re1 b6
13. a3 Bb7
14. Bg3 Rc8
15. Ba2 Bd6
16. d5 Nxd5
17. Nxd5 Bxg3
18. hxg3 exd5
19. Bxd5 Qf6
20. Qa4 Rfd8
21. Rcd1 Rd7
22. Qg4 Rcd8

Kasparov v. Karpov
World Championship, Moscow
1985-10-01

Figure 1: Procedural and explicit representations

Parameterization and constraint capabilities became important in CAD systems in the late 1980s, and are now available in all major CAD systems. Parameterization expresses design freedom, namely, what can be changed for design optimization, constraints express design restraints, namely, what must remain fixed to preserve design functionality. Both types of information specify *behaviour* of the model in the receiving system. Parameterization and constraints are *implicit* in a procedural model, and *explicit* in a B-rep model.

STEP, as published, cannot currently capture either of them. The secondary B-rep model can only be transferred by current STEP. This is just a 'snapshot' of an evolving model at some point in time. It lacks all data from the sending system concerning:

- Constructional history of the model
- Parameterization (design freedom)
- Geometric constraints (design restraint)
- Design features (high-level constructs with possible links to functionality).

This data is referred to as a part of the 'design intent'. Without it, the transferred model cannot be effectively edited after the transfer (it is a 'dumb model').

STEP enhancements must handle both procedural and B-rep approaches because systems routinely generate *hybrid* models in which some operands are explicit geometric models such as sweeping of an explicit 2D sketch to create a volume.

Therefore, the aim of the ISO TC194/SC4 Parametrics project is to enhance ISO 10303 for the transfer of design intent information. Consequently, to enable the exchange of CAD models which behave in the receiving system as though they had been created there. In particular, to transfer models that can be effectively edited or modified in the receiving system.

Procedural modelling will be discussed in clause 2 with special focus on the technical approach we adopted for the representation of procedural model using existing STEP deliverables, which has good harmonization with existing standards. In clause 3 to clause 7, major contents and status of standards we have developed will be explained, which are ISO 10303-55, 108, 109, 1101, 1102, 111, 112. The combined use of these standards for parametric shape model exchange is summarized in clause 8. Various pilot projects on parametric shape model exchange in U.S. and E.U. are summarized in clause 9.

2 Procedural modelling

2.1 Principles

A procedural model is defined as sequences of constructional operations. In the purest approach, no explicit faces, edges or vertices are included. But, most CAD systems support a hybrid model (hybrid procedural and B-rep model), where explicit entities may also be involved (e.g. (procedural) sweep operation on explicitly defined 2D profile).
ISO 10303-55 provides the necessary sequencing mechanism. It is intended that a procedural or history-based model will always be transferred together with a corresponding explicit model known as the *current result*. The current result is a fully evaluated B-rep model created by the sending system. A history-based model is inherently parametric. The current result is therefore a representative example of the family of models defined by the history-based model. The current result may be referred to as a check on the validity of the transfer, and possibly for the resolution of ambiguities.

2.2 Technical approach

In ISO 10303-55, it is intended to use EXPRESS shape modelling entities as constructional operations as explained below. ISO 10303-55 provides means to represent the constructional operations as a structured history file to cope with subdivision of design. Design rationale information can be included in each structure as text. It makes provision for history-based representation of non-shape models as well.

2.3 EXPRESS and Part21

EXPRESS representation of the class of circles is:
ENTITY circle
SUBTYPE OF (conic);
radius : positive_length_measure;
END_ENTITY;
Part21 (exchange file) instance of two particular circles is:

#420 = AXIS2_PLACEMENT_3D(...);
#425 = CIRCLE('C1', #420, 12.0);
#430 = CIRCLE('C2', #420, 15.0);

In an AP203 exchange file each instance defines an element of the model that is being transferred, element-by-element, from System-A to System-B.

We propose to use the same set of constructional entities and the same Part21 instance format for the transfer of the procedural model. But they are differently interpreted by the receiving system. When the model is procedural, the instance is interpreted as an instruction to System-B to create an instance of the specified entity type.

In an AP203 exchange file the order of the instances does not matter. They can be assembled in any sequence. In a procedural model, the sequence of constructional operations is extremely important. The same set of operations, differently ordered, usually gives a different shape model. For this reason, we collect instances that are to be interpreted as constructional operations into instances of **procedural_shape_representation_sequence**. EXPRESS representation slightly simplified is:

ENTITY procedural_shape_representation_sequence
    SUBTYPE OF (geometric_representation_item);
    elements : LIST[1:?] OF geometric_representation_item;
END_ENTITY;

Part21(exchange file) instance of a procedural_shape_representation_sequence is;
#420 = AXIS2_PLACEMENT_3D(...);
#425 = CIRCLE('C1', #420, 12.0);
#430 = CIRCLE('C2', #420, 15.0);
#435 = PROCEDURAL_SHAPE_REPRESENTATION_SEQUENCE
    ('PSRS1', #425, #430)

2.4 Advantages of the proposed approach

It is upwardly compatible with existing STEP parts. It allows construction of procedural models, immediately, in terms of any entities for which EXPRESS definitions exist in the standard. In particular, constructional representations can be transferred for models of non-geometric objects such as plans, processes, organizations, etc.

3 ISO 10303-55: Procedural and hybrid representation

3.1 Overview and status

This standard is a fundamental resource for procedural or constructional history modelling which is very different from other STEP modelling methods. Nevertheless, it is interoperable with other parts of STEP. It uses EXPRESS shape modelling entities in alternative role as constructional operations. A structured history file with provision for inclusion of design rationale information is enabled. It makes provision for history-based representation of non-shape models as well.

The DIS(Draft International Standard) document was 100% approved to publish as IS (International Standard) as a result of DIS ballot which was closed in May 2004. Its IS document is currently being prepared. It is intended that Part55 will be used by Edition 2 of AP203, which will provide component part representations in terms of feature-based construction history.

3.2 Schemas

The document contains two schemas, which are theprocedural_model_schema and the procedural_shape_model_schema.

The scope of the procedural_model_schema is:

- The specification of sequences of constructional operations for the procedural representation of models of any type.
- The hierarchical structuring of operation sequences.
- Hybrid models involving operations on explicitly defined elements.
- The use of entity instances as creation operations.
- The association of a procedural model with the explicit result of its evaluation.
- The capture of design rationale information in textual form.
- The capture of elements selected from the screen in the originating system.
- The suppression of operations for purposes of model simplification.

The scope of the procedural_shape_model_schema is:
- The specialization of the preceding schema for the particular case of shape models.

- Further specialization to procedural models generating specific types of explicit shape model as defined in ISO 10303-42 such as solid models, surface models and wire frame models.

The scope of ISO 10303-55 excludes:
- Persistent naming capabilities for the identification of model elements. The requirement of robust maintenance of reference is substituted by the specification to capture elements selected from the screen in the originating system.
- Conditional operation sequences involving IF THEN ELSE and similar constructs.

4 ISO 10303-108: Parameterization and constraints for explicit geometric product models

4.1 Overview and status

This standard provides representations for variable parameters (e.g. dimensional parameters), constraints and 2D profiles or sketches. Parameters and constraints are typically used to control:

- Geometric elements of 2D sketches
- Inter-feature relationships in 3D models
- Inter-part relationships in assemblies.

The DIS document was 100% approved to publish as IS as a result of DIS ballot which was closed in September 2003. Final document was sent to ISO for IS publication in June 2004.

4.2 Schemas

The document contains five schemas, which are parameterization_schema, explicit_ _constraint_schema, variational_representation_schema, explicit_geometric_ _constraint_schema and sketch_schema.

The scope of the parameterization_schema is:
- Association of variable parameters with quantities in an instanced EXPRESS representation.

- Specification of domains of validity for such parameters.
- Specification of entity instance attributes whose values are fixed.

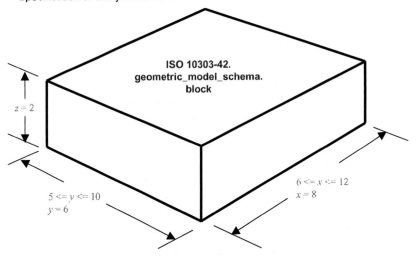

Figure 2: parameterization_schema example

The scope of the explicit_constraint_schema is:
- Specification of mathematical constraint relationships between parameters in a model.
- Specification of descriptive 'defined constraints' whose semantics are shared by the sending and receiving systems.
- Specification of simultaneous constraint groups, that may be hierarchically nested.

The scope of the variational_representation_schema is:
- Specification of a variational_representation as a model containing parameterization and constraint information.
- Association of a variational_representation with a fixed model (the *current result*), corresponding to current values of all parameters, and providing a representative example of the family of models defined by the variational_representation.

Status of the ISO TC184/SC4 Parametrics Standards Development

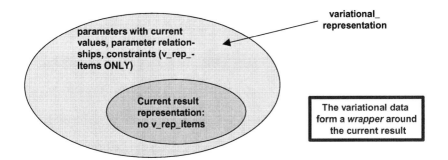

Items belonging to the current result also belong to the v_rep
Items belonging *only* to the v_rep are variational_rep_items
(parameters, constraints etc.)

Figure 3: Principle of variational_representation

The scope of the explicit_geometric_constraint_schema is:

• Specification of a range of explicit geometric constraints as descriptive relationships among elements in a product shape model, in either 2 or 3 dimensions. Examples are parallelism, perpendicularity, tangency, etc.

The scope of the sketch_schema is:

• Specification of sketches as planar geometric constructs for use in modelling operations such as extrusion.
• Specification of constraints among sketch elements.
• Specification of constraints between sketch elements and other elements external to the sketch.

The scope of ISO 10303-108 excludes:

• Representation of mathematical expressions and functions. ISO 10303-50 and ISO 13584-20 are used for this.
• Solution methods for constraint systems.
• Sequential application of constraints.
• Implicitly defined constraints.

4.3 General comments

The parameterization and constraint mechanisms are very general, and have application to any type of model. In addition to shape models, it may be applicable to processes, plans, organizations, properties, etc. Part108 is upwardly compatible with existing APs. It sets pointers into an AP203 type B-rep shape model but demands no modification of existing schemas to provide reverse pointers. Part108 parameterization and constraint data are redundant at the time of transfer. They come into play after the transfer, when they govern the behaviour of the exchanged model in the receiving system.

5 ISO 10303-109: Kinematic and geometric constraints for assembly models

5.1 Overview and status

Currently, STEP assembly modelling (based on ISO 10303-44) is of the parts list / bill of materials variety for the sets of positioned and oriented parts.

This standard aims to capture:

- Logical relationships between parts expressed as assembly feature relationships
- Kinematic degrees of freedom between parts.
- Parameterized assemblies, with inter-part constraints.

This will allow the full modelling of assemblies with parameterization and constraints.

The DIS document was 100% approved to publish as IS as a result of DIS ballot which was closed in November 2003. The final document was sent to ISO for IS publication in June 2004. Associated are two modules which had been developed together with Part109:

 - ISO 10303-1101: Product property feature definition

 -ISO 10303-1102: Assembly feature definition.

They were also agreed to publish as TS(Technical Specification) and were sent to ISO for publication at the same timing in June 2004.

The assembly feature definitions are compatible with APs 214/224, but also provide more direct access to underlying feature geometry. Constraint relationships are based on Part108. Part109 contains new concepts, and the modules contain necessary specialization of previously defined concepts.

5.2 Schemas

The part109 document contains two schemas which are assembly_feature_relationship_schema and assembly_constraint_schema.

The scope of the assembly_feature_relationship_schema is:

• Association of high-level assembly feature definitions with their detailed low-level geometric representations.

• Specification of relationships between associated features belonging to different part models.

The scope of the assembly_constraint_schema is:

• Specification of 3D explicit geometric constraints for use in modelling constraint relationships between components in assemblies.

• Specification of fixed component to anchor an assembly in space.

6 ISO 10303-111: 'Construction history features'

6.1 Overview and status

This standard is a resource for 3D feature based modelling in EXPRESS entity form.
Modelling mechanisms and feature representations were validated in the PDES inc. CHAPS project, which terminated November 2003. It is a new resource for Edition 2 of AP203.

It uses an approach for modelling features that is different from that taken in existing feature based STEP APs 214/224. It allows more direct access to feature geometry. It requires Part55 as its underlying resource.

The related document passed its combined NWI/CD ballot in December 2003 and its DIS version is under development.

6.2 Schemas

The document contains a single schema, construction_history_features_schema.
Feature types defined include the following:

- Rounds, fillets and chamfers
- Pockets (rectangular, circular, general)
- Slots, grooves (with various cross sections)
- Holes (with various bottom conditions, counter bores, countersinks)
- Bosses, pads and general protrusions
- Patternsof features(rectangular and circular arrays)

- Solid features defined by extrusions and revolutions of faces
- Solid features defined by thickening faces or shells of faces
- Thin-walled shapes defined by hollowing out solids('shelling')
- Shapes defined by cutting a solid with a general surface.

7 ISO 10303-112: Procedural models of 2D sketches

This standard is a resource construct for the neutral representation of sketch modelling commands. This activity is based on the results of transfer experiments performed at the Korea Advanced Institute of Science & Technology (KAIST). Capabilities to be included are based on a survey of 5 major CAD systems and experiences in the exchange of the database log file in neutral form.
Harmonization with other ISO 10303 resources is currently in progress.
NWI proposal was accepted and Committee Draft (CD) is under development.

8 How to represent parametric shape model by combined use of these standards

How to use these standards for parametric data exchange is as outlined below.
Parametric modelling history itself can be represented by 10303-55 as sequences of EXPRESS entity data type creation instructions. It allows structured representation of history and attachment of design rationale information at each structure in textual form. Shape elements picked from the screen can be represented by the use of 'selected_element' specification. Partial suppression of operations can also be represented.
Representation of a sketch as a combination of geometric elements and geometric constraints is enabled by 10303-108. It allows definition of variables, valid range of variables, assignment of current values, representation of algebraic constraints among variables.
As for inter-part geometric constraints necessary in assembly design, 10303-109 includes all the necessary constraints as specializations of 10303-108 specifications.
Though 10303-55 allows the appearance of any EXPRESS entity data type creation instructions, more specific modelling functions such as fillet, chamfer, shelling, etc. will be required for actual parametric data exchange. 10303-111 is intended to neu-

trally define these feature based 3D modelling functions in EXPRESS entity data types. Therefore, sending system's modelling command shall be transformed to corresponding function prepared in 10303-111, and then it shall be transformed to receiving system's corresponding modelling command.

If the creation of a sketch from scratch is also required to be exchanged, 10303-112 can be used in the similar situation as 10303-111.

9 Tests of new Parametrics capabilities

The first successful PDES Inc. transfer of a procedural model object was achieved in 2001(Theorem Solutions: CATIA vs. UG). This used WDs of new STEP resources. The work was continued in the PDES Inc. CHAPS project.

NIST had earlier demonstrated the transfer of a simple procedural model (SolidWorks vs. Pro/E) based on a 'non-standard' EXPRESS model. This work is being continued to bring it into line with ISO 10303.

KAIST in Korea is also working in this area., using a slightly different approach, with five CAD systems(CATIA, I-DEAS, Pro/E, SolidWorks, UG).

The university of St.Galen (Switzerland) has been working with ProSTEP and the German automotive industry in using ISO 10303-108 for the transfer of parameterized manufacturing information.

The most definitive test has been in the PDES Inc. CHAPS program, a 16-month effort that terminated in November 2003. The CHAPS program was funded by the US Office of Naval Research under its 'Supply Chain Practices' scheme. Six engineering companies were involved including Northrop Grumman, Raytheon and four of their suppliers. The CAD systems used were CATIA V4, Pro/E and UG. The translation software was developed by the British company Theorem Solutions Ltd.

The program verified the viability of the ISO 10303 approach to the exchange of constructional history models. It used development versions of ISO 10303-55, 108, 111 and 203ed2. Part models exchanged were of real parts suggested by participating engineering companies. The major problem encountered were due to the presence on some of the test parts of feature types outside the scope originally defined for the project.

A final written report on the CHAPS program is available. It demonstrated the successful transfer of 67% of the chosen set of test models between three CAD systems, about two-thirds of them without human intervention and the remainder after slight rework. This success rate is higher than the project goal (50%) and is a good rate for a pre-beta translator. The production version is expected to drastically raise the success rate.

Feedback from the project will be used to improve ISO 10303-55, 108, 111 and 203ed2 of the standard before their final publication.

It was found that prime contractors in general prefer to sacrifice some of the advantages cited above by transferring CAD models to their suppliers in the pure constructional history form without the associated parameterization and constraint data. The parametric information is regarded as commercially sensitive, and too revealing of design methodology. On the other hand, for company-internal model transfers, where confidentiality issues do not arise, the advantages of transferring the parametric information together with the constructional history are recognized

10 Conclusions

- The ISO TC184/SC4/WG12 Parametrics Group is extending STEP for the exchange of dynamic models. Its aims are:
    - ŏStandardization of procedural model representation in the new resource **ISO 10303-55**, currently being prepared for publication as IS.
    - ŏStandardization of parameterization and geometric constraints for explicit models in **ISO 10303-108**, sent to ISO for publication as IS in June 2004.
    - ŏEnhancement of assembly representation in **ISO 10303-109** and its associated two modules, **ISO 10303-1101** and **ISO 10303-1102**. These three standards were also sent to ISO for publication as IS for 109 and as TS for 1101 and 1102 in June 2004.
    - ŏStandardization of feature-based procedural (history-based) model representation using new **ISO 10303-111** resource for construction history features, of which DIS level document is currently being prepared.
    - ŏDevelopment of a new resource for the construction history of 2D sketches, **ISO 10303-112**, of which CD document is currently being prepared.
- Key concepts have been subjected to practical tests.
- Group collaborates with PDES Inc. and other organizations in this work.

11 References

[1] Mike Pratt, "ISO/IS 10303-108: Parameterization and constraints for explicit geometric product models", 2004
[2] Akihiko Ohtaka, "ISO/IS 10303-109: Kinematic and geometric constraints for assembly models", 2004
[3] Mike Pratt, "ISO/DIS 10303-55: Procedural and hybrid representation", 2004
[4] Bill Anderson, "ISO/DIS 10303-111: Construction history features", 2004
[5] Soonhung Han, "ISO/WD 10303-112: Procedural models of 2D sketches", 2004
[6] Michael Stiteler, "CHAPS Program Final report", 2004

12 Author

Akihiko Ohtaka, Nihon Unisys, Ltd., 1-1-1 Toyosu, Koto-ku Tokyo 135-8560 Japan
Email: akihiko.Ohtaka@unisys.co.jp

Living Vehicles – The Paradigm Shift towards Holistic Product Creation in the Automotive Industry

Jivka Ovtcharova

Institute for Applied Computer Science in Mechanical Engineering (RPK), Karlsruhe University (TH), Germany

Abstract: To achieve a real transformation of the traditional product creation in the automotive industry from a technology-centred towards a knowledge-driven approach, a new working environment is needed interrelating business, people, process and technology in a common network of activities throughout the overall product lifecycle. This consideration leads to a holistic product creation, which requires a new way of systemic thinking as opposed to just continuously and incrementally improving the technology side of the current development practice. This paper outlines a novel approach towards a holistic product creation considering a vehicle as a *living product*, which, as in nature, comes into existence through *genetic material or DNA*, is *born*, *grows* and reaches the necessary *maturity* for production, it stays *alive* when delivered to the customer and will *die* at its end of life. Accordingly, the main challenge for research and development is to establish new methods of work for a continuous integration of customer requirements with market trends, business objectives, people knowledge, process and technology. This approach requires the development and use of intelligent vehicle concept templates for capturing knowledge through vehicle lifecycle, as well as for tracking and ensuring customer satisfaction before start of serial production.

„The significant problems we face cannot be solved at the same level of thinking we were at when we created them".

Albert Einstein

1 Introduction

Current development in the automotive industry is challenged by a growing complexity of products and processes, while drastically reducing time-to-market and costs is viewed as one of the key competitive factors, e.g. [OvKl-03]. To increase their competitiveness in the global marketplace, major automotive companies develop business strategies that focus on the use of advanced information technologies to reach a higher degree of process automation. However, practical experiences have shown that information technology itself can not lead to a real *quantum change* to achieve a decisively new level of product development performance. The solution to be provided must be related to the establishment of a novel environment of work, shifting the *centre of gravity* from technology itself towards its integration with the business, human knowledge, and the processes. This consideration underlines the importance of the interaction of the four main factors - business, people, process and technology - as a network of activities with a common purpose, e.g., the development of a vehicle. It is the interaction and not one of the four factors alone that provides the key to success. The alignment of business, people, process, and technology means a paradigm shift of the overall product creation towards a holistic approach, i.e., unified understanding of the particular in the context of the whole. In accordance with [Weis-00], "...holistic is defined in two ways: the emphasis of the importance of the whole and the interdependence of its parts, and the concernment with wholes rather than the analysis or separation into parts."

Although there is a big amount of work to be done, breakthrough considerations towards a holistic product creation take already place in research, i.e. [ISTP-04] and industry, i.e. [Goul-03], in order to be able to envision new ways of operating business and strengthen advantages in future.

This paper outlines the scope and objectives of a novel approach towards a holistic product creation based on the consideration of a vehicle as a *living product*. Adopting a biological perception of vehicle creation the goal of this approach is to achieve significant improvement of quality and reduction of product development time through continuous integration and validation of the vehicle concept created through *genetic material or DNA*, (see also [MaOv-03]). The paper is organized as follows: Section 2 outlines the motivation and objectives of adapting a biological perception of vehicle creation. In Section 3, the underlying methodology is shortly described. Finally, Section 4 summarizes the status and advantages of the presented approach.

2 Motivation and Objectives

Radical advances in development and manufacturing of industrial products have been igniting efficiency explosions ever since Adam Smith introduced the division and specialization of labour, followed by Eli Whitney's idea of producing standardized and interchangeable machine parts or Thomas Edison's idea of performing product development by a dedicated research and development group. These *quantum breakthroughs* are all examples of engineering solutions which pushes industry to a new life competition leap. Even more recent quantum improvements of this nature have included the advent of virtual development, product lifecycle management, e-business, and the high-speed/low-cost *mass customisation*.

In accordance to [Grat-99], a *quantum breakthrough* occurs when an organisation implements a decisively new working and management process that enables it to achieve an entirely new level of development performance. It can graphically be characterized by a *process-performance curve,* which is almost always *S-shaped* as shown in Figure 1. When a new process (red curve) is initially implemented, performance improves slowly first, since there is an inertia (confusion, resistance, and a learning curve to surmount). Performance then accelerates rapidly as the new process gains momentum, generates irrefutable results, wins converts, and becomes widely adopted. The quantum change creates an entirely new process performance curve that is discontinuous from the current one (blue curve). On the other hand, using current process and technology still makes *incremental change* possible, however without the capability of a vastly higher performance improvement.

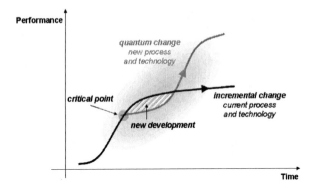

Figure 1: Quantum Process Change (according to [Grat-99])

The automotive industry is nowadays on the verge of a similar process curve. It is time for a new quantum change. The vehicle has transformed from being just a means for transportation to something that can provide passengers with options for enhanced comfort, entertainment, safety, and even a mobile office with phone, fax, and Internet connections. Rising customer needs and wishes require automotive companies to design new vehicles that create *value-added experiences* tailored to each customer. Customer loyalty is quickly becoming elusive for most of the automotive manufacturers as they can now gain or lose customers with a mouse click Armed with better information and greater market power, customers are demanding superior quality vehicles that meet their specific, often unique, requirements.

To improve competitiveness and ensure customer satisfaction, a decisively new working environment for automotive development has to be developed, which, as opposed to the current state, is human-centred and not just technology-oriented, as shown in Figure 2.

Living Vehicles – The Paradigm Shift towards Holistic Product Creation

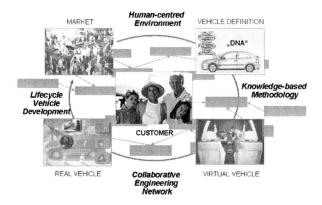

Figure 2: New Working Environment

Transitioning to the new working environment as outlined above requires the following scientific and technological objectives being fulfilled:

- Establish a *Human-centred Environment* for enabling a real collaboration of humans (customers and workers involved in the vehicle creation process) to better understand customer wishes and support collaborative thinking, working, and training, as belonging to one community.

- Develop a *Knowledge-based Methodology* for continuously integrated product development based on a *vehicle DNA* definition, an intelligent engineering template, as well as on collaborative design and validation activities.

- Create a *Collaborative Engineering Network* using web technology as a platform for connecting business, people, processes and technology in real-time.

- Enable *Lifecycle Vehicle Development* across dynamic, virtual organisations, to support continuous integration along the entire value chain, geared toward delivering customised solutions in a timely and cost-effective manner.

In the following section, the methodology for *living products* is briefly outlined.

3 Methodology Outline

The key idea of the proposed approach is that vehicles are considered *living products,* which like in nature (see Figure 3), are created using genetic material or DNA, grow and reach the necessary completeness and maturity of high-level organisms, ready to be produced and delivered to the customers. Therefore, a living vehicle possesses a biological genetic history and can be born with desired properties defined by a general design pattern, i.e., vehicle DNA that carries out the instructions of making living products from one generation to the next. In accordance to the biological perception, the genetic vehicle history implies all the primary items well known from the past development, plus new, recent requirements (see Figure 4).

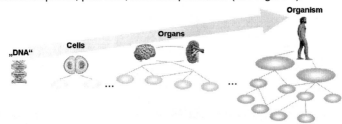

Figure 3: Harmonic Creation Process in the Nature (High-level Organism)

To enable a harmonious growth process in the vehicle development, a network of interrelating building components and systems has to be established which provides transparency to all designers and workers involved into the product creation process. Furthermore, the whole product lifecycle should be taken into consideration. Following the main lifecycle phases of *living organisms,* the proposed approach considers the following four interrelated lifecycle phases of a vehicle: *Life, Premise, Genesis and Growth.*

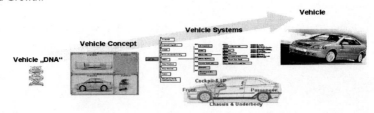

Figure 4: Harmonic Creation Process of Vehicles

Living Vehicles – The Paradigm Shift towards Holistic Product Creation

In the following, the four vehicle lifecycle phases are briefly described (see Figure 5)

Life - production and maintenance of current vehicle generation

The development of new generation vehicles always starts with the study of customer wishes and satisfaction, as well as with the analysis of current generation vehicles. During this phase, serial production, use, maintenance and continuous improvement are taking place and important perceptions and statistics concerning the next generation vehicles are gained. The objectives in this phase are to support continuous improvement of vehicle and strengthen relationships with the customers. In particular, customers' broadband connections at home and work, combined with emerging Internet services could help product development teams to reach customers more quickly and less expensively ensuring a fast and accurate acquisition of requirements. In this phase, strong emphasis should be put on the definition of customer profiles. Who exactly represents the major group of customers? What are their requirements, wishes and needs? Due to the heterogeneous customer community, the requirements need to be obtained from diverse information sources. Once available, this information is used in the *premise* phase.

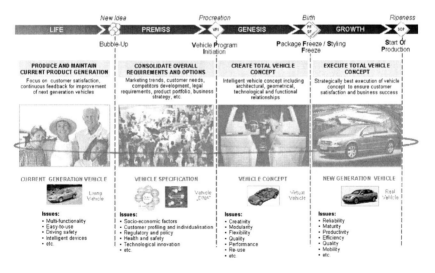

Figure 5: Living Vehicle Creation Phase

Premise - **consolidation of requirements and options towards next generation vehicles**

Based on the existing *best practices* of vehicle development as well as the customers' input from the *life* phase of current generation vehicles, the main idea here is to use a biological-like DNA structure, containing desired properties. This phase describes mainly the preparation work towards a new generation vehicles. The objectives are to ensure best consolidation of overall requirements and model options including market trends, customer needs, and competitors' trends analysis besides legal requirements, product portfolio, business strategy, etc.

Following the biological perception of harmonic creation process, it is not enough to simply merge existing methods and specifications into one document. The *vehicle DNA* should be considered as a generic engineering template from the beginning of a vehicle program, containing information on both, the main building parts, components and systems of a vehicle and their relationships. Modifying the generic code in a kind of evolutionary process leads to the brand specific product. From a biological point of view, this is comparable to the human evolution in the early stages of mankind to humans nowadays as shown in Figure 6.

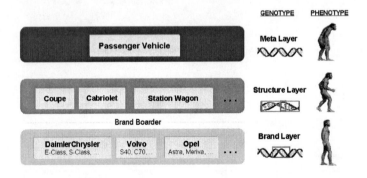

Figure 6: Vehicle Meta Scheme

Here, a transparent product data structure is needed to facilitate necessary product modifications and to radically decrease the workload of any modernisation or customisation of the product. As a consequence, the productivity of designers could increase and the time needed for product development will be significantly reduced.

Advantages of this approach are the better *organic/genetic* consolidation of various factors such as socio-economic factors, customer profiling and individualisation, regulatory and policy, safety and mobility.

Genesis – generation of new vehicle concepts

This phase characterises the procreation of the new vehicle, based on the *genetic DNA code*, developed in the *premise* phase. Here, the rapid evaluation of ideas and concepts early in the development process and the identification of important *delighter features* are of high importance. The objectives of this phase are to ensure a strategically best executable vehicle concept based on architectural, technological, functional as well as physical associability. In particular, the functional view will enable modelling and assessment of functions and sub-functions with their respective properties, the physical view enables the use of information technology to show interdependencies, assess appearance and check e.g. movements, the process view provides criteria to measure the concept against market, customer, manufacturability etc. [Ovtc-02, Ovtc-04]. Advantages of this approach could be the better support for design, integrity, modularity, flexibility and efficiency as well as the higher performance and quality of work, the increase of knowledge re-use and the collaborative thinking and working.

Growth – execution of the new vehicle concepts till begin of serial production

This phase characterises the real growth and ripening process of the new vehicle generation. Based on the vehicle concept developed in the *genesis* phase, the vehicle is now developed in details and validated against different criteria. The so called *organic* product development is finished at this stage. The objectives here are to ensure strategically best producible and saleable vehicle based on virtual validation of all criteria to guarantee customer satisfaction and business success.

Summarizing the approach introduced in this paper, along the vehicle lifecycle development, main stress was put on the *premise* and the *genesis* phases, which can be characterised as *development-intensive*. Here, a breakthrough innovation will be reached through a consolidated *vehicle DNA definition* and an intelligent vehicle concept. Not less important but dependent on the two development-centred phases are the *life* and *growth* phases. These can be characterised as *service-intensive* with clearly assured breakthrough innovation through the direct integration of acquisition and management of customer satisfaction as well as tracking and maintenance of customer requirements to ensure customer satisfaction before start of serial production.

4 Conclusion

> „It is not the strongest species or not the most intelligent one that will survive. Those will survive who are reconfigurable to changes".
>
> Charles Robert Darwin

This paper outlined the scope and objectives of the *living products* paradigm adapted to the vehicle creation process and introduced a methodology based on the consideration of four main vehicle lifecycle phases: *life phase* of the production and maintainance of current vehicle generation, *premise phase* of consolidation of overall requirements and options concerning the next generation vehicles, *genesis phase* of creating the new vehicle concept, and a *growth phase* for executing the new vehicle concept until serial production.

The key inovations and benefis of the outlined approach are among others:

- Transformation of traditional *technology-oriented* towards a *knowledge-driven* vehicle development

- Support for a *customer-oriented development* environment to fulfil the requirements of a brand management, portfolio planning (demand and economically feasible product orientation), vehicle concept engineering and vehicle integration, customer satisfaction measures input into product definition, as well as detailed interface description for product development execution

- Support for a *lifecycle vehicle development* across dynamic virtual organisations, including supplier collaboration

Future development will focus on extending and refining the proposed research within the scope of an industrial project.

5 References

[Grat-99] M. McGrath, *Looking for the Next Big Thing*, PRTM's Insight Magazine, December 1999.

[ISTP-04] IST Priority on *Products and Service Engineering in 2010*, EC/FP6, 2004. http://www.cordis.lu/ist/so/engineering2010/events/wp_excerpts.htm

[MaOv-03] R. Mayer-Bachmann, J. Ovtcharova, *Living products: Biologische Erkenntnisse für die industrielle Wertschöpfung*, CAD-CAM Report Nr.12, 2003.

[OvKl-03] J. Ovtcharova, K. Kleinmann, *Collaboration in Virtual Engineering. Industrial Business Case: Automotive Development*, Proceedings of the e-Challenges 2003 Conference, Bologna, Italy, 22-24 October 2003.

[Ovtc-02] J. Ovtcharova, *Virtual Vehicle Development at Adam Opel AG - Industrial Case Study*, Proceedings of the 7. Seminario International de Alta Technologia, UNIMAP, Piracicaba, Brasil, 9-11 September, 2002.

[Ovtc-04] J. Ovtcharova, *Virtual Engineering – Herausforderung und Chance am Beispiel der Automobilindustrie*, J. Gausemeier, M. Grafe, AR & VR in der Produktentstehung, HNI-Verlagsreihe, Band 149, Juni 2004.

[Weis-00] U. Weissflog, *Holistic Virtuality – The Need for Balance of People, Process and Technology in the Emerging Digital World*, ProduktDatenJournal, Nr.2, 2000.

[Goul-02] L. Gould, *PLM: Where Product Meets Process*, Automotive Design & Production, 2003. http://www.autofieldguide.com/articles/060203.html

6 Author

Jivka Ovtcharova, Institute for Applied Computer Science in Mechanical Engineering, Universität Karlsruhe, Kaiserstrasse 12, D-76131 Karlsruhe, Germany

e-mail: ovtcharova@rpk.uni-karlsruhe.de

Geometric Modeling of Layout Constraints for Plant PLM *(or New Challenges for Solid Modeling imposed by PLM)*

Nickolas S. Sapidis

University of the Aegean, Ermoupolis, Greece

Abstract: Robust Product Lifecycle Management (PLM) technology requires availability of informationally-complete models for all parts of a design-project including *spatial constraints*. (The specific application considered by this author is "industrial-plant design", yet the findings of this research may be applied in many areas of product design.) This is the subject of the present investigation, leading to a new model for spatial constraints, the "virtual solid", which has been analyzed in a series of papers; see [TS'04] and references therein. The present publication focuses on the solid-modeling aspects of the virtual-solid methodology, and derives new solid-modeling problems (related to object definition and to object processing) that appear in layout-constraint analysis. Solutions are proposed, which should be integrated in the solid-modeling kernel that supports spatial-constraint modeling/processing.

1 Introduction

The author has been following recent developments in the *"PLM/PDM/Collaborative-Design"* areas (and also has been contributing some related research) and has concluded that technical and commercial publications, in this exciting field, are suppressing the relevance of geometric modeling to these important technologies. Indeed, a typical research-paper or software-brochure on PLM or PDM will talk a lot about *information/data modeling & management, product structure, relations between various entities, international standards for data exchange, computer-network technologies, XML, data bases,* but very little about geometry! We all understand that this is the obvious strategy for the PLM/PDM community to differentiate itself from the CAD community, which indeed has been overemphasizing, for many years, the importance of geometry in product design. However, this approach leads to the impression that all these new developments create no new problems/challenges for geometric modeling, which is simply *wrong!*

This research (see also [T'03], [TS'04] and references therein) has been focusing on an integrated PLM model for complex plants, putting emphasis on both the geometric and the nongeometric aspects of the related problems. The present paper deals with the former; the latter part is detailed in [T'03] and [TS'04]. Specifically, the main part of this work (Sections 2 and 3) studies real-life 3D layout-constraints and establishes that these are best modeled/analyzed within an "extended solid-modeling environment" offering state-of-the-art solid-modeling as well as solutions to new geometric problems arising directly from typical 3D layout-constraints.

2 A Geometric Modeling Methodology for Spatial Constraints based on "Virtual Solids"

The specific research-proposal, of representing spatial constraints as "extended solids" (or "virtual solids") using solid-modeling technology, was introduced in [ST'99] and subsequently further investigated in a series of papers (see [TS'04] and references therein) and in the PhD Thesis [T'03]. Prior to this work, similar ideas, in a much more primitive form, appeared in few publications in the fields of CAD, computer graphics and robotics. E.g., Chang and Li [CL'95] study 'assembly maintainability' employing user-defined "constraint volumes". These are modeled using spheres blended with each other by frustums of cones. Kim and Gossard [KG'91] solve a 'packaging task' with a method, which is a typical example of the 'optimization methodology' (Section 2.2.2). Yet, the introductory part of the paper includes a comment on the 'accessibility problem', proposing the creation of 'a fictitious solid object with the shape of the desired access space'. This idea is not explored any further in this or any other paper of these authors.

2.1 Modeling Layout Constraints using "Virtual Solids"

A mechanical product includes a number of components whose position must satisfy various design constraints. [SI'97] considers the related 2D design problem, and classifies these constraints into three groups: *dimensional constraints, regional constraints, and interference constraints.*

The first group includes distance and angular dimensions, like those appearing in engineering drawings, that constrain the size, geometric form, location and orientation of components. Regional constraints restrict the region (area) where a component may lie, and finally, interference constraints specify components and regions that must not overlap. Current methods to solving these constraints adopt the following fundamental strategy [ST'00]:

- *All constraints are translated (exactly or approximately) into dimensional constraints.*
- *Dimensional constraints are viewed as either a system of algebraic (in)equalities or as a graph, and a solution is produced by a numeric or symbolic or graph-theoretic method.*

This methodology is rightly characterized as ``design with dimensional constraints'' and is one of the very early approaches investigated by CAD pioneers, more than twenty years ago. It is based on the fundamental assumption, implied by standard engineering drawings, that

"a design is fully described by a prescribed number of parameters".   (*)

This assumption is true only for arrangements including very simple geometric objects with trivial topology. Thus, it is not a surprise to anyone that despite the research efforts of twenty years --- which considerably advanced the available tools for treating constraints --- the scope of the method is still limited to 2D and trivial 3D problems. Even for 2D problems, traditional "design with dimensional constraints" has many disadvantages:

a. The underlying assumption (*) is, in general, erroneous and should be applied only to 2D layouts with simple geometric objects.

b. The produced techniques often lead to ``black box'' software-tools, i.e., they allow no user involvement.

c. These methods are unable to take advantage of evolving solid modeling and CAD technologies.

Our proposal, "solidified constraints", implies exactly the inverse approach [ST'00][TS'04]:

(i) All design constraints are translated into properties (or constraints/problems) of real and/or virtual solids and relationships between these solids.

(ii) The resulting "virtual-solid modeling problems" are treated by procedures operating on these solids.

Obviously, regional and interference constraints may be directly described as ``solidified constraints'' according to (i), thus we focus on dimensional constraints. Those describing features of a solid may be directly incorporated into the explicit or parametric description of the solid. In the case a parametric solid modeler is not available, then one faces the problem of incorporating inequality constraints like ``*the height H of the object X must be less than h_0*''. In this case, one must use virtual solids and translate the dimensional constraint into a regional one. For the above example, the geometric description of **X** and the value h_0 uniquely define a virtual solid X_0 allowing statement of the above constraint as ``*X must be a subset of X_0*''. Finally, dimensional constraints describing relations between components are incorporated into the solid modeling system using either virtual solids or appropriate local coordinate systems.

Analyzing real-life industrial examples – see many cases in [T'03] and [TS'04] -- leads to the conclusion that the task of classifying/modeling a design-constraint is not a trivial procedure and it is doubtful that this can be fully automated. This is demonstrated by the example of Fig. 1.

Geometric Modeling of Layout Constraints for Plant PLM

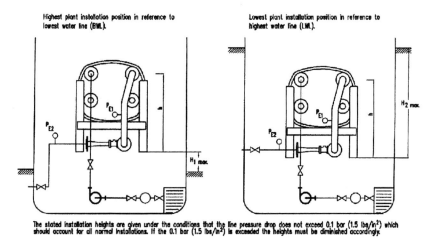

Figure 1: Highest and lowest plant position with respect to standard horizontal-planes.

The two constraints defined in Fig. 1 are "in principle" dimensional constraints as they refer to the position, in the vertical sense, of the depicted plant-system. However, it is straight-forward to see that these constraints are essentially regional constraints and can be modeled, using 2D or 3D virtual solids and set-operations, as explained in Fig. 2.

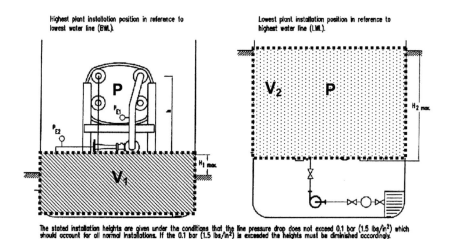

Figure 2: [Left] Highest plant-position constraint is equivalent to: $P \cap V_1 \neq \emptyset$, and [Right] lowest plant-position constraint is equivalent to: $P \subseteq V_2$.

137

2.2 Component Layout using "Solidified Constraints"

It must be clarified that our approach does not aim at eliminating dimensional constraints from the design process but rather placing them at a correct perspective in view of today's object-oriented solid modelers. We propose viewing these as constraints on properties of (real or virtual) solids and processing them in a solid-modeling environment, to derive improved design-tools. This approach allows "constraint solvers" to take full advantage of the rapidly evolving CAD and solid-modeling technology as opposed to traditional "design with constraints" techniques that view layout problems only as systems of equations, and thus rely primarily on analytic, graph-theoretic and numerical methods. Indeed, a huge variety of approaches to ``design with constraints" is available, where the differentiating factors are: "amount of interactiveness", "scope of constraint set", "level of compartmentization", "level of optimization", etc (see discussion and references in [T'03]). The two extreme cases ("fully interactive" and "fully automatic" layout), and their relevance to "solidified constraints", are discussed briefly below.

2.2.1 Interactive Component Layout

Very seldom an industrial designer adopts a global approach to constraint solving, simply because the corresponding tools operate in a ``black box" mode, i.e., the only input is the constraints, and in a single step, a (usually ``unique" and/or ``optimal") solution is produced, to which the designer sees often obvious improvements. For this reason, designers prefer the interactive approach where, the design problem is subdivided into smaller sub-problems, which are treated individually in an interactive manner. This means that, to each design modification, the software must respond with a "report" on constraints satisfied/violated. This "report" must be carefully designed so that it offers maximum support to the designer's work without overwhelming him/her with "esoteric" geometric information. Here, solidified constraints are clearly superior to standard constraint modeling as the latter can only inform the user about violations of constraints. Employing "solidified constraints" would allow use of any current solid modeling system that may offer valuable new functionalities, like: (i) appropriate measures for "closeness to constraint violation", (ii) recommendations about minimal design-variations that would lead to a valid arrangement, (iii) intuitive methods to bias a solution using ``constraint volumes" [CL'95], etc. Furthermore, treating constraints in a solid-modeling environment guarantees that the system can handle any number of constraints for arbitrarily complex geometric objects (e.g., with free-form spline surfaces). Indeed, the research highlighted here has been implemented in the AutoCAD system, and can handle even solids with free-form faces as well as nonmanifolds, as AutoCAD offers a robust environment for 3D solids (based on the kernel ACIS) and also for 2D "solids" (AutoCAD calls these ``regions").

2.2.2 Global Optimization of a Layout

In this approach, all design constraints are assembled in a system of (in)equalities treated by an optimization procedure that operates on the whole set of design parameters and produces a solution also minimizing an appropriate ``cost function". This is the classical version of the ``design with parameters" approach, discussed above, which, in principle, is incompatible with the new proposal of ``solidified constraints". However, if the constraint solver can collaborate with external procedures (today's solvers can definitely do that), then it is possible that materializing certain constraints as solid-modeling queries may improve the scope and the robustness of a "global optimization" method. Indeed, [T'03] evaluates in-detail current techniques to handle 2D regional constraints, and it establishes that these are not as robust as the corresponding solid-modeling tools. A second example would be any problem involving a triangulation of a set of points, e.g., the ``Delaunay triangulation" [SP'03]. Here, construction of a valid triangulation requires robust and efficient application of the "point-in-circle(sphere)" test classifying a point as IN/ON/OUT with respect to a circular disk (or sphere). Clearly, this is best solved by the corresponding 2D or 3D solid-modeling function operating on a point and a primitive solid (circle or sphere) [SP'03].

3 Modeling Layout Constraints using "Virtual Solids": Requirements Imposed onto the Solid Modeling (SM) Foundation

The first research efforts, in the field of "geometric modeling of design constraints", have been relying on primitive-solids (box, cylinder, etc.) and elementary SM techniques (extrusions and set operations) to describe spatial constraints for maintainability analysis [CL'95] [ST'99]. Subsequent work (see [TS'04] and references therein) has revealed that industrial applications involve also spatial constraints that impose severe requirements on the employed SM kernel, as, e.g., they involve some new SM problems [ST'00]. Here is a partial list of these requirements:

3.1 Special Requirement "A": *2D & 3D Solid Sweep*

3D Solid-Sweep is involved in many cases of spatial constraints; see Fig. 3. Although the "general solid-sweep", with an arbitrary 3D path, is still an unsolved problem, various special cases are solvable [T'03]. These are the cases of solid-sweep most often involved in free-space modeling: e.g., in Fig. 3[Left] the related solid-sweep problem is primarily 2D, while in Fig. 3[Right] the solid-sweep problem is equivalent to rotating a surface around an axis. Both these problems have exact solutions (these are detailed in [T'03]), which should be included in the SM kernel supporting free-space design/analysis.

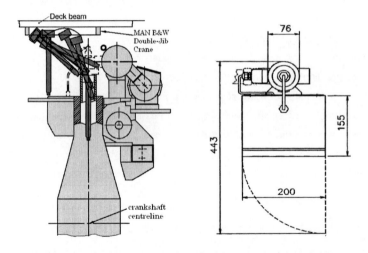

Figure 3: [Left] Tilted lift of a ship-engine's piston using a double-jib crane: the required free-space is defined by a «general solid-sweep». [Right] Opening the cover of a UV sterilizer's control unit: the required free-space is defined by rotating this cover around an axis.

3.2 Special Requirement "B": *Robust 2D/3D Solid Simplification*

A vital component of a free-space modeling system is the "2D/3D object simplification procedure", as this system employs plant components (e.g., see Fig. 4) and/or structural elements (e.g., see Fig. 5) which are defined by final fully-detailed drawings or CAD-models (e.g., "production drawings/models" or "PLM models").

Using these as "input information" often leads to over-detailed SM descriptions of spatial constraints. Indeed, in Fig. 4: This engine's «Lubricating Oil Cooler» (the component at the low-left end of Fig. 4(a) shown also in Fig. 4(b)) corresponds to the required free-space VS_C. Definition of this volume involves faces of the «Cooler», which must be simplified (prior to construction of VS_C) so that VS_C is free of unnecessary details.

In the example of Fig. 5, the "exact free space" (Fig. 5[Left]) is a virtual solid with numerous faces corresponding to a SAT file of 67 KB (exported from an AutoCAD implementation of this research [TS'04]).

However, the given layout-problem changes negligibly if one uses the "approximate or simplified free space" of Fig. 5[Right], which includes only five faces and is described by a SAT file of 5 KB (exported from AutoCAD), reducing the amount of required memory-space by 93%!

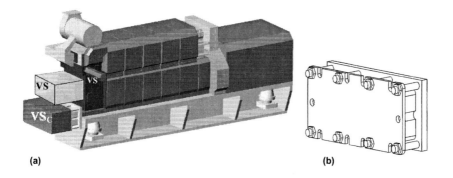

(a) (b)

Figure 4: An engine and its required free-spaces.

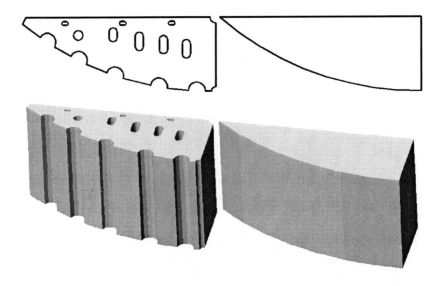

Figure 5: [Left] A volume defined by extruding a given profile. [Right] The «same» volume by extruding a simplification of that profile.

It is noted that current «2D simplification» methods are, in general, not appropriate for the present application, as they cannot easily accommodate the requirements of free-space management [ST'99]. Indeed, in [ST'99] we have developed a new «2D simplification» for the needs of the present problem.

It must be noted that, although this last algorithm is quite compatible with the requirements of free-space modeling, it is still far from fully serving all needs of the present application: e.g., in the example of Fig. 5, the simplified profile (Fig. 5[Right]) should be allowed to extend beyond the convex-hull of the original profile while the algorithm [ST'99] forces simplified regions to lie within the convex hull of the original region.

3.3 Special Requirement "C": *Modeling Mixed-Dimensionality Unbounded Volumes*

Often a manufacturer or designer describes required free-spaces (RFSs) for humans and/or equipment by specifying required "floor area" or "traffic lanes" or by (partially of fully) describing volumes using surfaces; see Fig. 6.

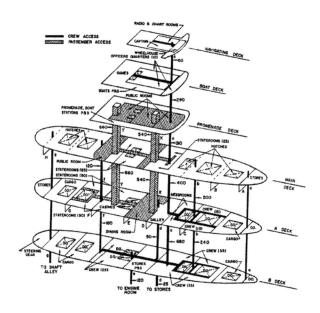

Figure 6: Access diagram for crew and passengers in a ship.

Here, one deals with "volumetric objects" that may (temporarily or permanently) violate several of the fundamental laws of SM. For instance, such an object (i) may be only partially defined by one or more planes/surfaces or profiles, (ii) may be unbounded in certain directions (e.g., see the two "vertical volumes" in Fig. 6), (iii) may include a variety of geometric entities (e.g., wireframe models or just lines), some of which (are supposed to) describe "symbolically" a volume, and (iv) may use *nonstandard geometric operations* to define required free-spaces.

3.4 Special Requirement "D": *Simultaneous Handling of Fully- and Partially-Evaluated Free Spaces*

When a designer deals with the geometric definition of a product/assembly/system, he/she always fully evaluates the specified entities so that he/she can visually inspect the design, apply interrogation/analysis procedures, communicate his/her proposals to others, etc. This is not, in general, the preferred practice when one designs required free-spaces: here, one needs only to make sure that (a) the whole design (= "product models" + "design constraints") is valid, and (b) that the RFS definitions are included, in some appropriate form, in the design database .Thus, RFSs may be unevaluated solid models, a practice that leads to a very economical description of these entities. E.g., the typical RFS is an extruded object defined by a planar closed polyline and a "height-value" (e.g., see Fig. 7(a)). Clearly, storing only this polyline and the real number defining the "height-value" is much more economical than storing the resulting extruded solid (Fig. 7(b)).

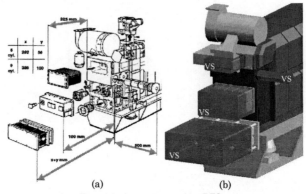

Figure 7: *An engine and its RFSs.*

Using "fully unevaluated" SMs to define RFSs is not a realistic approach, as some of the RFSs, e.g., the most complex ones or those corresponding to "crucial areas" of the design, will have to be "detailed" as explicit 3D objects by the designer. Thus, in general, there always will be RFSs that must be described in a (fully or partially) explicit manner. This directly leads to the conclusion that the SM environment used for design/analysis of RFSs must be very robust and flexible regarding the creation/editing/combination of 2D/3D solids with a varying degree of "explicitness" in their definition.

3.5 Special Requirement "E": *Extending Assembly-Modeling Subsystem to Include Free Spaces*

It is straightforward to realize that that the obvious way to combine RFS information with the corresponding electromechanical system is to employ "assembly modeling" methodologies. Existing assembly-modeling methods mimic standard mechanical-assembly design where relationships between components are well-defined in terms of "mating conditions" ("against", "fit", "contact", etc). Often, the mating condition between an RFS and a mechanical part is a standard mating condition; this is true for the example in Fig. 8(a), where a face of the depicted RFS and a face of the engine are related with an "against" condition. However, sometimes no standard mating condition can describe the exact relationship between an RFS and one or more solids/assemblies; see, e.g., Fig. 8(b), where no face of the RFS may be associated, with a standard mating condition, to any face of the two engine-assemblies.

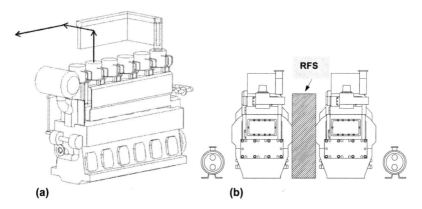

Figure 8: (a) A «generalized assembly» including a standard assembly (ship engine) and a virtual solid (RFS for piston removal). (b) A «generalized assembly» including a virtual solid (RFS for maintenance personnel) and two standard assemblies (ship auxiliary engines).

Extending the list of mating conditions is not the only required modification of the "assembly modeling subsystem" for it to be able to handle arbitrary RFSs. As explained above, often an RFS is only partially defined by one or more "surfaces", which also must be included in an "extended assembly" (= parts + constraints + RFSs) as "2D virtual solids"; see example and discussion in Fig. 9.

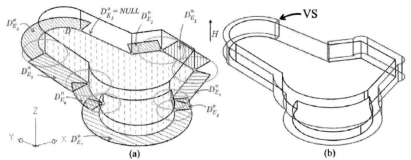

Figure 9: (a) A «generalized assembly» including a standard solid and several 2D virtual solids representing «required floor areas». (b) The RFSs implied by the «required floor areas».

3.6 Special Requirement "F": *Interactive Editing/Handling of Non-Manifold Geometry*

Modern geometric-modeling systems do represent and handle nonmanifold geometry as this is often involved in geometric constructions. E.g., a trivial sequence of regularized set-operations may produce, as an intermediate result a nonmanifold object, despite the fact that the final outcome is a manifold solid (see example in Fig. 10).

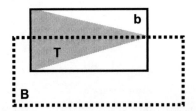

Figure 10: Although the final outcome of the Boolean expression *«(b – T) ∩ B»* is a very plain manifold-object (the orthogonal triangle shown, in this figure, exactly under the triangle T), the intermediate result *«(b – T)»* is nonmanifold.

Geometric Modeling of Layout Constraints for Plant PLM

However, no current CAD system allows interactive manipulation of nonmanifold features, simply because one can think of no practical situations where a designer/engineer will need to define such a geometric configuration.

Surprisingly, exactly the opposite is true when a designer operates on certain kinds of RFSs appearing often in practical situations. E.g., given a solid **R** and a plane of reference **Π** (e.g., corresponding to <z=0>), the volume **R-** under **R** (down to the plane **Π**) is a typical case of RFS [TS'04]. The algorithm calculating a solid-modeling description of **R-** [ST'00] constructs simultaneously (many) parts of the three volumes **R-**, **R+-** and **R+** (: lying under/in-the-shadow/above **R**, respectively), which, in general, may touch each other at nonmanifold vertices and/or edges.

Thus, the final step of the procedure relies on user-involvement who must «cut» appropriate nonmanifold connections to identify all parts of the volume **R-**. (The algorithm in [ST'00] does include a method automating this «nonmanifold cutting» operation. However, this is not sufficiently robust for complex real-life situations.)

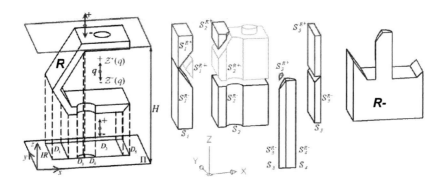

Figure 11: [Left] A solid R and a plane of reference Π. [Right] Construction of the volume **R-** under **R** (down to the plane **Π**) involves processing of nonmanifold volumes.

4 Conclusions

This research has been focusing on identifying and solving new geometric-modeling problems related to the development of an informationally-complete model for layout-constraints. This model would be useful in many CAD applications, including the development of an efficient PLM/PDM system for complex plants. The "strategic objective" of this work is to emphasize that the evolving PLM/PDM methodology is bound to give rise to new geometric problems that cannot be solved by existing solid-modeling tools. A detailed analysis has been presented identifying crucial functionalities that must be included in a SM kernel supporting design of spatial constraints.

Although most of the current PLM/PDM literature (commercial or research-oriented) continues to suppress geometric aspects of this exciting new technology, it is very encouraging to see industrial PLM/Collaborative-Engineering initiatives, like "JT Open" (http://www.jtopen.com), putting sufficient emphasis on the "solid-modeling side" of PLM.

5 Acknowledgements

The content of this paper has been significantly improved through comments expressed by Prof. C. Werner Dankwort and participants of the 5th International Workshop on Current Cax Problems; the author thanks them all.

This research builds on results produced by the author and Dr. G. Theodosiou; the author acknowledges many fruitful discussions with G. Theodosiou during the last eight years as well as his many suggestions and comments.

This work was supported by the General Secretariat for Research & Technology of the Ministry of Development (Greece) through the Operational Programme "Information Society" (E-business Project 26 "E-MERIT: An Integrated E-Collaborative Environment for Product & Process Modeling Using 3D Models and Avatars", 2003-2005) and through the Bilateral Greece–France S&T Co-operation Programme (Project "Information Management and Exchange in Network-Centric Product Design: Modeling Component Knowledge in Design Repositories", 2003-2005).

This research has also been supported by the Research Centre ELKEDE – Technology & Design Centre" (Athens, www.elkede.gr) through a number of R&D grants.

6 References

[CL'95] H. Chang, T.Y. Li, "Assembly Maintainability Study with Motion Planning", Proc. IEEE International Conference on Robotics & Automation, 1995, pp. 1012-1019.

[KG'91] J.J. Kim, D.C. Gossard, "Reasoning on the Location of Components for Assembly Packaging", Trans. of ASME: J. Mechanical Design, Vol. 113, 1991, pp. 402-407.

[SP'93] N. Sapidis, R. Perucchio, "Combining Recursive Spatial Decompositions and Domain Delaunay Tetrahedrizations for Meshing Arbitrarily Shaped Curved Solid Models", Comp. Meth. in Appl. Mech. & Eng., Vol. 108, No. 3, 1993, pp. 281-302.

[ST'99] N. Sapidis, G. Theodosiou, "Planar Domain Simplification for Modeling Virtual-Solids in Plant and Machinery Layout", CAD, Vol. 31, 1999, pp. 597-610.

[ST'00] N. Sapidis, G. Theodosiou, "Informationally Complete Product Models of Complex Arrangements for Simulation-Based Engineering: Modeling Design Constraints using Virtual Solids", Engineering with Computers, Vol. 16, 2000, pp. 147-161.

[SI'97] H. Suzuki, T. Ito, H. Ando, K. Kikkawa, F. Kimura, "Solving Regional Constraints in Components Layout Design based on Geometric Gadgets", AIEDAM, Vol. 11, 1997, pp. 343-353.

[T'03] G. Theodosiou, "Solid Modeling of Free Spaces in Plant Layout: Application in Design of Ship's Engine Room", Ph.D. Thesis, Depart. of Naval Architecture & Marine Engineering, National Technical University of Athens, 2003.

[TS'04] G. Theodosiou, N. Sapidis, "Information Models of Layout Constraints for Product Life-Cycle Management: A Solid-Modeling Approach", CAD, Vol. 36, 2004, pp. 549-564.

7 Author

Nickolas S. Sapidis, UNIVERSITY OF THE AEGEAN, Department of Product and Systems Design Engineering, Ermoupolis, Syros, GR-84100, Greece,

sapidis@aegean.gr, www.syros.aegean.gr/users/sapidis/

The "Digital Factory"

Franz-Josef Schneider

University of Applied Science, Stuttgart, Germany

Abstract: The article describes scope, goals and status of implementation of the "Digital Factory" initiatives from car manufactures, which is from the IT perspective currently one of the most innovative projects in automotive industry.

1 Introduction

The "Digital Factory" includes and manages all digital data and corresponding IT systems which are necessary to develop and build a product (e.g. car). The "Digital Factory" provides the IT environment for concurrent engineering processes of product and manufacturing disciplines. All IT applications will be taken under the umbrella of the "Digital Factory". To improve quality of planning, to provide planning results in an understandable manner virtual reality (VR) methods are used.

This initiative requires a network of data management systems (EDM systems) and packages of CAD, simulation and VR tools which work closely hand in hand without any breaks in database and workflow.

2 Scope and goals of the "Digital Factory"

The "Digital Factory" includes digital data from product engineering, process planning and manufacturing engineering. The connector between product and manufacturing engineering is the manufacturing process. (see figure 1).

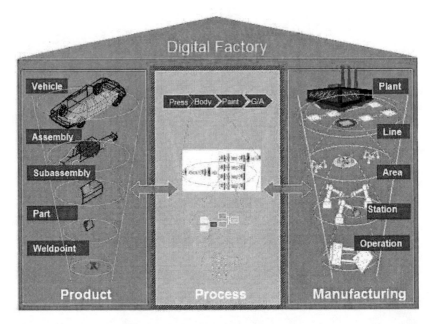

Figure 1: Scope of the "Digital Factory".

The time before the "Digital Factory" was established product engineering and manufacturing engineering were separated from IT perspective. The manufacturing process was planned only with little IT support like Excel-spreadsheets. In planning phases engineers lost most of their time searching for actual planning data or using non actual data from other planning departments. So goals of the "The Digital Factory" are:

- reduce planning time
    - get quick access to actual data
    - upgrade concurrent engineering

- improve planning quality
    - evaluate digital models
    - use common visualisation platform to discuss planning results.

The "Digital Factory"

3 Implementation of the "Digital Factory"

In order to link product engineering to manufacturing engineering and to get really a benefit from digital planning methods two major columns must be set up:
- EDM network
- VR system

3.1 Implementation of EDM network

The first column of the "Digital Factory" is the implementation of a network of EDM systems, which work closely together. To set up a strong workflow between different planning departments this common used network is mandatory. Main tasks for the EDM network are:
- provide uniform environment for navigation, searching for and storage of data
- share important data between different applications of engineers
- maintain data consistency
- inform the user at once when digital models were changed from an application
- fit VR tool with all necessary data.

Figure 2: EDM Network of the "Digital Factory".

3.1.1 Current status of implementation:

- currently different data management systems for product engineering and manufacturing engineering are in use
- the GUI, navigation and storage of data will be handled completely different in systems
- no online data sharing, only sharing via export, import between EDM systems is possible
- no EDM network is in place, which takes all important manufacturing data and applications under its umbrella
- data feeding of VR tool are based on file import.

3.2 Implementation of a VR system

The second column of the "Digital Factory" is a common used VR tool. The tool deals as an discussion platform. Planning results from different departments and disciplines are joined together to get the "big picture". Planning results can be visualized very fast and in an understandable manner, also for non experts. So more people like workers or management can be involved into early planning phases and can give valuable input.

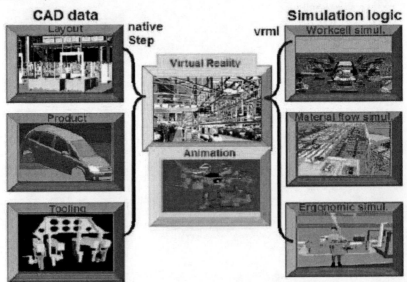

Figure 3: VR tool of the "Digital Factory".

The "Digital Factory"

The main tasks for the VR tool are:

- visualization and evaluation of digital models

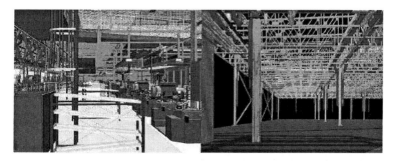

Figure 4: Navigate inside digital models.

- checking and discussing planning results (e.g. collision check).

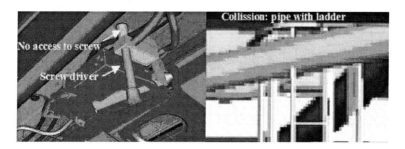

Figure 5: Collision checks.

3.2.1 Current status of implementation:

- data import in native or step format for geometric models are used but rework must be done
- data import of simulation logic and the connection to geometric objects (models) is not possible
- animation as an alternative instead real time simulation can be done, but additional expenditure is necessary
- handling of the huge amount of viewing data is not satisfactorily, no automatically generation of adapted level of details (LOD) for different specific animation scenarios is in place.

3.2.2 Future use of VR system

In future VR methods will be used directly in plants to overlay actual real data with digital planning data. Checks of planning scenarios will be done on location.

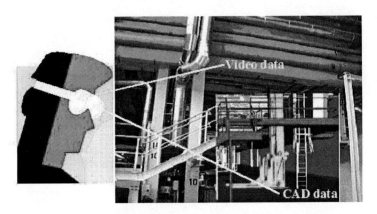

Figure 5: Mobile VR tools.

4 References

The results of this article are based on my work at the "Adam Opel AG" from 1999 to 2004 as a member of the "Virtual Factory" group and my collaboration with other German car manufactures as a member of the VDA working group "Digitale Fabrik" from 2001 up to now.

5 Author

Prof. Dr.-Ing. Franz-Josef Schneider, Hochschule für Technik Stuttgart/ HFT-Stuttgart
Schellingstr.24, 70174Stuttgart
franz-josef.schneider@hft-stuttgart.de

Convergence Engineering

Hiromasa Suzuki

The University of Tokyo, Tokyo, JAPAN

Abstract: In the Multi-Media world "Convergence" has the specific meaning that all the conventional media such as books, paintings, photos, music records and movies will be represented in digital forms so that they can be easily edited, transmitted, retrieved and archived. In the field of manufacturing, similar things have happened. By the advancement of X-ray CT technologies, real objects in manufacturing can be digitized into computers and integrated with digital engineering systems. Such real objects include products, tools, materials and so on. The most important concern here is to fully exploit digitized data for developing new digital engineering processes beyond today's CAX systems by incorporating objects and manufacturing processes in the real world into digital world. We refer to such a way of engineering as "Convergence Engineering." For this purpose, it is crucial to import digitized data smoothly from the X-ray CT into the CAX systems. However, today it requires tedious manual work in this process. In this paper, we introduce our recent research results for generating meshes from CT volume models which are more directly fed to CAX systems.

1 Introduction

By recent advancements of X-ray CT technologies it becomes a standard task in some company to scan mechanical parts and to generate digital models of the parts [2]. Among those advancements, improvement of scanning speed reduces the scanning time so that it becomes possible to scan the whole body. And the improvement of scanning accuracy allows the scanned data to be used for reconstructing 3D models. Figure 1 shows an example of a CT image for an cylinder head of an engine and figure 2 shows a volume model built by piling up such sectional images as shown in figure 1.

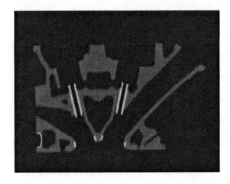

Figure 1: X-ray CT image of a section of a cylinder head (courtesy of Toyota Motor Corp).

Figure 2: Volume model for a cylinder head (courtesy of Toyota Motor Corp).

Though the most fundamental function of X-ray CT is "non-destructive imaging of internal structure," it is now recognized as "3D scanning device" to generate 3D image (volume model) and furthermore 3D surface (mesh) models as well. And such scanned 3D information has been used in various applications. One application is shape comparison to compare the mesh model of a real object with its 3D CAD models. By evaluating the difference and dimensional variation between the CAD model and its produced part, we can improve the quality of product design and manufacturing processes. Furthermore, we can compare different real objects to know the variation under different manufacturing conditions. We can also compare the shape of a product before and after a manufacturing process so that we can trace the change of the shape of a particular part through the manufacturing process. It allows us to analyze effects of manufacturing conditions systematically so that we can reduce the trial and error lead time needed for manufacturing preparation.

Such scanned 3D models are further used for CAE simulation such as casting simulation and air flow simulation. It is reported that such simulation using real object's shapes match better to experimental results than using CAD models.

As you can see from those applications, it is a key how far the scanned data is utilized in design and manufacturing processes. What I call "Convergence Engineering" is a new concept for digital engineering where the scanned 3D models are fully utilized combined with CAX applications. In Convergence Engineering physical objects are converged with digital models for realizing innovative design and manufacturing methodologies.

2 Iso-Surfacing and Its Problems

Therefore in Convergence Engineering System, it is fundamentally important to input scanned data to CAX systems. The direct output from the X-ray CT system is a volume model while the major representation of a CAX system is a solid/surface model. So we usually generate mesh models from the volume model; and those mesh models are further converted to be used in CAX systems. However this conversion requires tedious, time-consuming manual work. This is one of the biggest problems which hinders the Convergence Engineering, since data transfer from X-ray CT to CAX is crucial to the Convergence Engineering.

There are range of conversion processes including various problems. In this paper we focus on one important aspect of them related to the today's standard methods to generate mesh models. These mesh models are generated from the volume model by using iso-surfacing methods (contouring methods). The Marching Cubes algorithm [3] is the most famous method and most widely used. In those methods, the volume model is considered as a continuous implicit function $f(x, y, z)$ sampled at the three dimensional orthogonal regular grids. And iso-surfacing methods extract a surface $f(x, y, z) = \theta$ where θ is an iso-value or threshold value. So, if meshes needed by CAX systems or designers are not an iso-surface of the volume model, they can not be directly generated by those methods. In such cases, tedious, time-consuming manual work to correct the generated meshes is needed. We will introduce two of these problems in the following sections, thin plate structure and multi-material parts.

3 Mesh Reconstruction from X-Ray CT Data

3.1 Thin Plate Structure [1]

Let us consider the case of thin plate structure. Figure 3 (A) shows a typical example of sheet metal parts. Such thin plate structures are usually designed by their medial surface with uniform thickness and are represented as a surface model in CAX systems. However, in the scanned image of such a thin plate structure, the medial surface does not form an iso-surface. So it is not possible to apply the Marching Cubes directly. Figure 4 describes this problem in 2D. Figure 4 (A) depicts a volume model for the thin plate structure. By applying the Marching Cubes, we obtained a mesh as shown in Figure 4 (B). It generates a solid model even though the object is very thin. It is more desirable to generate medial surface models as shown in Figure 4 (C). Generating a medial surface mesh from the solid surface mesh is usually very difficult.

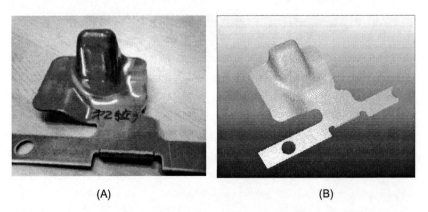

Figure 3: A thin plate structure (A) and its medial surface mesh (B).

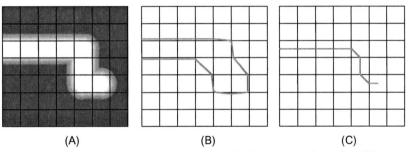

Figure 4: Iso-surfacing the thin plate structure (A, B) and its medial surface (C).

We developed a method to generate medial surface meshes from the volume model. Figure 5 shows an overview of our method. We first convert the grey level volume model into a binary model by thresholding (Figure 5, (A)). Second we apply so-called "skeletonization" method [4] to extract voxels at the medial surface (B). Third we define its polarity, that is a plus/minus value to indicate the front and back surface of the plate (C). Finally we apply marching cubes to extract a medial surface.

Figures 3 (B) shows an extracted medial surface mesh for a part shown in Figure 3 (A). The size of the volume model is 400x450x150 and the number of triangles of the mesh is about 150,000.

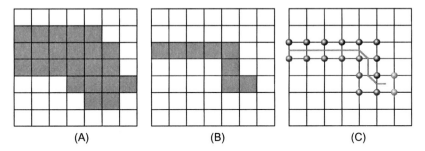

Figure 5: Medial surface extraction process. (A) a binarized volume model, (B) skeletonization and (C) an iso-surfacing with local polarities.

3.2 Multi-Material Parts

The next topic is to deal with a multi-material object. A part is often made of several materials. Figure 1 shows our typical example of a cylinder head of an automotive engine. Its major part is made of aluminium (shown in grey color) and it also includes some iron parts (shown in white color). So there are three materials of aluminium, iron and air. The problem is how to extract material boundaries as mesh models. In this case, there are three kinds of material boundaries, separating iron and aluminium, aluminium and air, and air and iron. Respectively there are three threshold values to distinguish two of those materials. We want to extract these boundaries as mesh models. However, if we apply a general Marching Cubes algorithm, we get problematic results, since we can specify only a single threshold value. Furthermore, we can not define a proper threshold value to the area where all the three materials meet. So we can not generate a mesh in this area using the Marching Cubes algorithm.

We propose a method to extract material boundaries by adapting threshold values depending on the material types and use extended version of Marching Cubes to generate meshes. We first perform segmentation to classify material types and their boundaries. Then we apply the Marching Cubes to extract material boundaries by adapting the threshold value depending on the material types. For the area where the three materials meet we adapt the voxels and polarities so that the Marching Cubes can consistently generate a mesh. We needed to extend Marching Cubes to handle multi-materials and also to deal with non-manifold meshes, because the multi-material boundaries have branches where more than three boundaries meet.

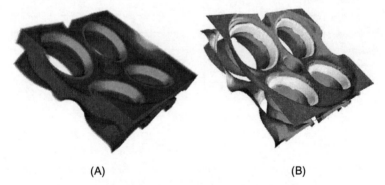

(A)   (B)

Figure 6: Material boundary extraction for multi-material parts. (A) volume model and (B) extracted non-manifold mesh.

Figure 6 shows an example. Figure 6 (A) shows a part of the volume model shown in figure 2. It is a top of a cylinder where intake/exhaust valves are attached. Figure 6 (B) is the result of mesh generation. The dark grey portion shows the aluminium surface while the light grey portions show the iron part. The size of the volume model is 80x200x170 and the number of triangles of the mesh is about 380,000. Figure 7 shows an enlarged view of a section of the portion where the three materials meet. You can see a branch structure of a non-manifold mesh. We also developed a mesh simplification method for non-manifold meshes by which we can reduce the number of triangles as much as you wish while keeping the original shape.

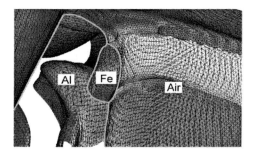

Figure 7: Enlarged sectional view of the multi-material part.

4 Summary

In this paper, we first introduce a concept of Convergence Engineering and discuss the importance of generating meshes which are more adaptive to CAX systems. The message here is that simple Marching Cubes is not enough. I introduced two applications, thin plate mesh generation and multi-material mesh generation. Using those mesh generation methods, the gap between the X-ray CT scan and CAX systems will be reduced.

Though X-ray CT is still an expensive investment for a company, the investment can be justified by Convergence Engineering technologies that improve the quality of products and lead time of product development. As the application areas of Convergence Engineering will be extended, we expect that more extensive research and development must be conducted.

5 Acknowledgements

I would like to thank Toyota Motor corp. for giving us their valuable comments and experimental data.

6 References

[1] Tomoyuki Fujimori, Hiromasa Suzuki, Yohei Kobayashi, and Kiwamu Kase. Contouring medial surface of thin plate structure using local marching cubes. In Proceedings of the International Conference on Shape Modeling and Applications 2004, IEEE Computer Society Press, 2004, pp. 297-306.

[2] Takahiro Okada, Yasuhiro Miwata and Hiroyuki Ishii, Three-Dimensional Shape Measurement with High-Energy X-Ray CT Scan, SAE 2003 World Congress, No. 2003-10-0133, 2003, pp. 1-5.

[3] William E. Lorensen and Harvey E. Cline. Marching cubes: A high resolution 3d surface construction algorithm. Proc. Computer graphics and interactive techniques, ACM Press, 1987, pp. 163-169.

[4] Steffen Prohaska and Hans-Christian Hege. Fast visualization of plane-like structures in voxel data. Proc. Visualization 2002. IEEE Computer Society, 2002, pp.29-36.

7 Author

Hiromasa Suzuki, RCAST, The University of Tokyo, Japan, 4-6-1, Komaba, Meguro, Tokyo 153-8904, Japan,

suzuki@den.rcast.u-tokyo.ac.jp

Modern Products: New Requirements on Engineering Design Elements, Development Processes, Supporting Tools and Designers

Christian Weber

Saarland University, Saarbrücken, Germany

Abstract: Modern products have become ever more complex to meet customers' and society's demands. At the same time, product development/design processes are expected to lead to better (and, in terms of variants, to more numerous) results in shorter time. This contribution names the challenges for future product development processes and tries to draw conclusions with regard to engineering design elements and related procedures, development methods and processes, supporting tools, and people involved. The aim is to initiate and contribute to a discussion between engineering design managers and practitioners, developers of supporting tools and researchers/teachers on ways to tackle present and future challenges. There are no turnkey solutions yet – and they probably cannot be provided based on one single strategy and/or realised by one single enterprise or institute.

1 Introduction

The main challenges for the development/design of complex "high-technology" products such as cars, air and space vehicles, machine tools, etc. are:

1. Stronger competition, global scale
2. More innovation required, development time ("time to market") decreasing
3. Risks (economic and technical) increasing
4. Number of variants increasing ("mass customisation")
5. Product complexity increasing
6. New forms of work distribution and organisation (e.g. supplier-integration, [computer-supported] co-operative work, …)

List items 1 to 3 name the dominant economic challenges, which product-creating companies have to master in order to stay competitive. Items 4 to 6 denominate technical challenges, which result from strategies that were introduced to cope with the economic requirements.

This contribution tries to investigate the impact/requirements of the challenges named above on

- engineering (design) elements,
- development methods and processes,
- supporting tools,
- people involved (developers/designers).

Before going into this discussion, a brief statement on the term "complexity" is presented, because a lot of current CAx-problems seem to be related to it.

No final solutions are presented, "only" some thoughts on how to meet the challenges of the future.

2 What Is "Complexity"?

Interestingly, the (mainly technical) challenges 4 to 6 named in Section 1 all have to do with "increased complexity" of products and product development processes. The term "complexity" or "complexity management" is often used in current discussions, but different aspects of complexity are often mixed up.

The author's concept of "complexity" is that in product development five different types of it have to be distinguished:

- "Variational complexity": Number of variants of a product (list item 4).
- "Numerical complexity": Number of components in a product (list item 5, product complexity, first aspect).
- "Relational/structural complexity": Number of relations and inter-dependencies between the components (list item 5, product complexity, second aspect).
- "Disciplinary complexity": Number of disciplines involved in creating the product (list item 5, product complexity, third aspect).
- "Organisational complexity": Distributed work, "global co-operation", "top-down" instead of "bottom-up" design, etc. (list item 6).

High-tech companies have increased all five types of complexity at the same time, and are dearly looking for methods and tools to support them coping with complexity (or rather: complexit*ies*).

The author assumes that the types of complexity proposed above require different supporting methods and tools. Additionally, it can be stated that current CAx-systems are mainly concerned with the variational and numerical types of complexity, sometimes, however, advertising their solutions to these two as solutions to complexity as such. Some initial (and increasing) activities address the handling of relational and organisational complexity types, but especially relational complexity is mostly seen too "part-oriented" and too "geometric" (utilising conventional BOM-strategies and CAD-parametrics) and ignoring the fact that many more types of relations than consists-of- and geometric relations exist between components and play a vital role in product development.

Little or no concepts exist of how to cope with disciplinary complexity. At the same time, disciplinary complexity (of products and processes) is one of the major challenges in present and even more so in future product development/design:

- Simultaneous/concurrent engineering strategies already today involve marketing, production planning, service experts, etc. into product development.
- The development of mechatronic products – and most of the formerly purely mechanical products are or will become mechatronic products in order to stay competitive! – needs the combination of mechanical design, hydraulic engineering, electrical/electronic engineering and software development.
- The next step will probably be the integrated modelling and integrated development of products consisting of material as well as immaterial components (so-called Product-Service Systems) [1, 2].

3 Engineering Design Elements and Related Procedures

As was discussed in Section 2, mastering "disciplinary complexity" is one of the major challenges of future product development processes. To a considerable extend, these processes rely on pre-defined elements and related procedures, which in a broader perspective can be seen as elementary "carriers of product and process knowledge".

When looking into traditional, i.e. "mono-disciplinary", concepts of defining engineering design elements and procedures, vast differences between disciplines can be detected:

- Mechanical engineering:
    - Engineering process dominated by design and calculation/simulation/optimisation *of elements* (e.g. machine elements, components of products).
    - Element/component structures of products even determine many company structures.
    - But: No clear taxonomy of elements (can be function-related, shape-related, technology-related, related to sourcing of components, ...).
    - No explicit functional considerations, physical aspects dominant.
    - Verification (calculation/simulation/optimisation) of element behaviour strong.
    - Geometry, tolerances, and surface quality very complex and important.
    - Manufacturing issues (of elements!) very important, **many technologies**.
    - "Interfaces"/dependencies between elements not systematically considered.
    - System synthesis/configuration and verification (calculation/simulation/optimisation) of system behaviour comparatively weak.

    Approach: mainly **"bottom-up"** (combining elements to systems, figure 1).

- Electronic engineering (similar: hydraulics/pneumatics):
    - Design **with elements**.
    - Elements pre-defined by functions and physical realisation; quite standardised taxonomy of elements and their behaviour, elements often identical with buyable components.
    - Verification (calculation/simulation/optimisation) of element behaviour weak (often not necessary).
    - Geometry, tolerances, surfaces no issue.
    - Manufacturing issues very weak (in electronics: **only one technology**!).
    - Interfaces/dependencies between elements thoroughly scrutinized.
    - Engineering process dominated by system configuration.

- Verification (calculation/simulation/optimisation) of system behaviour strong.

Approach: extremely *"bottom-up"* (combining elements to systems).

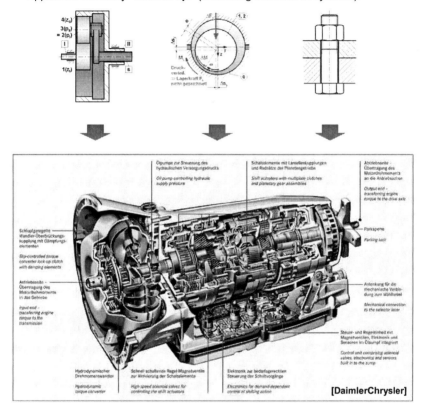

Figure 1: "Bottom-up" approaches prevailing in mechanical engineering

- Software engineering:
    - Abstract "elements" (algorithms, data-structures, ...), pure functional view.
    - Interfaces/dependencies between elements important, but difficult to capture.
    - Engineering process dominated by functional decomposition and modularisation.
    - Rapidly changing hardware platforms and software "paradigms".
    - Verification of software (systems) a big problem!

Approach: mainly *"top-down"* (decomposing system behaviour to element specification).
- Service engineering:
  - In science even treated in another faculty (economics, not engineering).
  - "Holistic" approaches, development methods/strategies very weak.
  - No systematic consideration of constituents of services ("service elements").
  - But: very interesting, customer-related evaluation concepts.

A more detailed investigation of differences between (mechanical) product development and service engineering can be found in [3].

For future product development processes, increasingly characterised by multi-disciplinary products, processes and organisations, engineering design elements and related procedures have to change in the following way:

- "Holistic", multi-disciplinary concept of engineering design elements (covering mechanical, hydraulic/pneumatic, electronic, and software engineering, maybe even including service components).
- Unified modelling approach.
- In order to enable easy system synthesis/configuration/simulation.
- Elements must contain information about
  - functionality/behaviour,
  - with compatible interfaces across disciplines,
  - tolerances (more than geometric tolerances),
  - defined dependencies between functionality/behaviour and design parameters (where appropriate),
  - manufacturing attributes (where appropriate).
- "Holistic", multi-disciplinary product development strategies with stronger focus on:
  - "Design **with** elements" instead of "design **of** elements" (as is prevailing in mechanical engineering).
  - Verification (calculation/simulation/optimisation) of system behaviour.
  - "Top-down" approach instead of "bottom-up" approach (again: as is prevailing in mechanical, but also in electrical engineering).

Modern Products: New Requirements on Engineering Design Elements,
Development Processes, Supporting Tools and Designers

In industry, the need for the changes outlined is clearly seen. The situation in engineering practice is characterised by the fact that related problems already have to be solved while industry itself as well as research is still looking for, developing and trying out appropriate concepts, procedures and organisations.

In research and teaching, some supporting activities can be noticed, although they are still relatively sparse. Examples for approaches and projects, which the author knows quite well and is partly involved in, are:

- Since 1997, a working group of German and Swiss university teachers (Arbeitskreis neue Lehre Maschinenelemente, AKME) is developing and applying new concepts and strategies for teaching of machine elements. Many of the requirements stated above are core concerns in these activities [4].

- Studying and optimising product development/design processes is a research topic originally brought up and still dominated by mechanical engineering (see, as examples, [4, 5, 6]). Here, a significant shift of paradigm can be observed right now which will be addressed in Section 4.

- The author is, at present, strongly engaged in developing advanced methods and tools for the modelling and simulation of mechanical components, based on so-called port-system concepts, which are already well known in electrical, hydraulic and control engineering. These activities, which are built upon basic considerations already presented some years ago [8, 9], can be seen as one contribution to the development of unified modelling and simulation concepts with strong focus on the functional behaviour of heterogeneous elements in multi-disciplinary system contexts.

  As an example, Figure 2 shows the block diagram of a gear wheel pair, which can be used in higher-level system block structures as well as to (automatically) generate the (differential) equations describing the functional behaviour of the component.

- Finally, activities addressing the integrated modelling and development of products consisting of material as well as immaterial components (Product-Service Systems) have already been mentioned [1, 2, 3].

The consequences for the development of supporting (CAx-) tools will be discussed in Section 5.

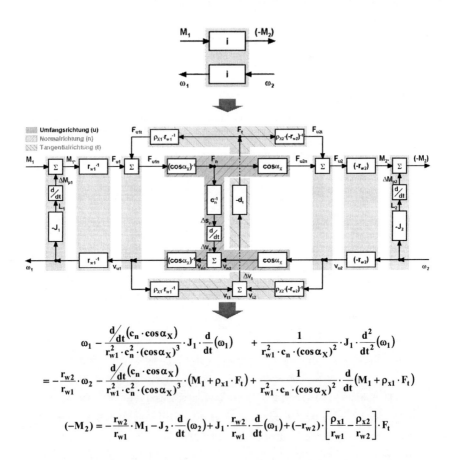

Figure 2: Modelling and simulation of mechanical components based on port-system concepts – a contribution to the development of unified and multi-disciplinary techniques

4 Development Processes

The main focus to adapt product development processes to present and future requirements again stems from the challenge of mastering "disciplinary complexity" (see Section 2). As was already outlined in the previous Section 3, the aim is:

- "Holistic", multi-disciplinary concept of product development processes,
- with strong focus on "top-down" (instead of "bottom-up") approach,
- and utilising unified engineering design elements (main topic of Section 3).

Additionally, new control mechanisms for development processes are required:

- Engineering processes controlled and driven by analysis/simulation results (Figure 3).

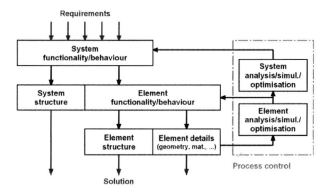

Figure 3: Flow scheme of "top-down" development process, controlled by analysis/simulation results

Present research activities to be mentioned in this context are:

- The very recent VDI-guideline 2206 on systematic development of mechatronic products [10] is an advanced and very interesting approach to extend product development methodologies as they are known for years in mechanical engineering (see e.g. [4, 5]) into a multi-disciplinary environment. The overall process approach defined in VDI 2206 is shown in Figure 4.

- Many research projects focus on new development methods for multi-disciplinary products, in Germany mainly sponsored by the national science foundation (Deutsche Forschungsgemeinschaft, DFG).

- The author is involved in developing a new approach to modelling products and product development processes (called "Characteristics-Properties Modelling", CPM, for products, and "Property-Driven Development", PDD, for processes). This approach is especially aimed at controlling/driving product development processes by analysis/simulation results and integrating existing methods and tools such as DFX ("Design for X", e.g. for manufacturing, for assembly, [11]), or CAx-components [12, 13] and is also the theoretical base of considerations on the integrated modelling and development of Product-Service Systems [2, 3].

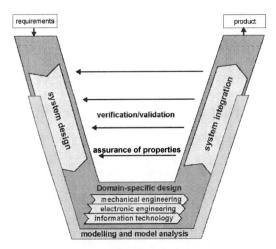

Figure 4: Scheme of product development methodology for mechatronic systems ("V-model") according to VDI 2206 [10]

5 Supporting Tools

Tools should always follow the processes they are intended to support – and not vice-versa (which, however, seems to happen in some cases with CAx-components ...). This simply means that the changes in development process concepts have to become the guiding principles for tool development in the field of CAx. Therefore, the issues discussed in Section 4 ("holistic", multi-disciplinary concept; focus on "top-down" approaches; utilising unified engineering design elements; processes controlled and driven by analysis/simulation results) could be repeated in this section as a general framework.

More specific conclusions for the development of CAx-tools are:

- Future product development processes will require even more computer support than we have today.
- Geometry and structure of products – the prevailing entities represented and processed by today's CAx-systems – address only one side of product modelling (in terms of "our" CPM approach: the "characteristics" of products, [11, 12, 13]).

- Additionally, (advanced or new?) CAx-tools must be able to represent and handle functionality/behaviour of products and their elements (in terms of the CPM approach: the "properties" of products, [11, 12, 13]).

  Reason: The required process control by analysis/simulation results can only be supported if appropriate CAx-components can capture, maybe interpret and process functionality/behaviour properties.

- The implementation of "top-down" approaches and the utilisation of unified engineering design elements (which, among other things, must have functionality/ behaviour attributes and defined interfaces, see Section 3) require extended ("parametric"?) concepts as a means to handle interfaces/dependencies between elements. (Developments into this direction are already visible in advanced CAx-systems.)

The author's opinion is that present PDM/PLM-systems are the most suitable base to develop them into core CAx-components for process control and integration in product development. To reach this aim, specific goals for the further development of PDM/PLM-systems are:

- "Parametric" PDM/PLM-systems required for the handling of interfaces/dependencies between elements and for the handling of dependencies between functionality/behaviour and structure (see above).
- "Multi-view" (and "multi-parametric"!) PDM/PLM required in order to handle many properties, which all play important roles in the product development process simultaneously (e.g.: functional structure ≠ parts structure ≠ manufacturing structure ≠ ...).

As was already stated before, a major part of the author's research activities are in the field of new architectures for PDM/PLM-systems, based on a new approach to modelling products and product development processes (CPM/PDD), which was also already mentioned [12, 13]. In co-operation with industry, these ideas are put on test in practical environments [14].

6 People Involved

The most important success factor for product development processes is motivating the people involved in the process to play an active role in the ongoing changes (and maybe also: to motivate enough young people to involve themselves into engineering at all!). The process-related issues discussed in Section 4 ("holistic", multi-disciplinary view; "top-down" approaches; unified engineering design elements; processes controlled and driven by analysis/simulation results) can be seen as a framework of these changes.

In general, a much broader qualification profile of development engineers will be required, because knowledge in many disciplines is necessary to cope with increased disciplinary, but also organisational complexity of products and processes.

Not all of the people can possibly have all of the broad *and* deep knowledge, however. Even if this thought must be regarded as speculation at the present time, it may be that – in initial as well as in further professional training – a distinction between two different profiles of product developers will have to be introduced:

- The system development engineer on one side,
    - having the "holistic", multi-disciplinary qualifications to synthesise system architectures,
    - break them down into element specifications, and
    - having the responsibility for the system-level analysis/simulation/optimisation loops (see Figure 3),
- and the element development engineer
    - with specialised qualifications to transform specifications derived from the system-level into element solutions,
    - having the responsibility to run the element-level analysis/simulation/optimisation loops, and
    - providing the information necessary for system analysis/simulation/optimisation.

The common base of co-operation of the two types of developers must be the "top-down" process approach, driven by analysis/simulation/optimisation loops on the two levels indicated.

Activities in research and teaching that deal with the investigation of future qualification profiles and the development of (initial or further) training programmes are, at present, relatively sparse.

Many changes in teaching have been initiated in the last couple of years, e.g. by the AKME-group of university teachers working on the renovation of machine elements teaching (as mentioned at the end of Section 3), but they may have to become more rapid and maybe a little more open to more radical changes. Additionally, the issue of further training of practitioners should get a much stronger focus than is the case today.

7 References

[1] Bullinger, H.-J.: Dienstleistungen für das 21. Jahrhundert – Trends, Visionen und Perspektiven. In: Bullinger, H.-J. (ed.): Dienstleistungen für das 21. Jahrhundert., Stuttgart 1997, pp. 27-64
[2] Weber, C.; Pohl, M.; Steinbach, M.; Botta, C.: Diskussion der Probleme bei der integrierten Betrachtung von Sach- und Dienstleistungen. In: Proceedings of the 13. DfX-Symposium, Universität Erlangen-Nürnberg, 2002, pp. 61-70
[3] Weber, C.; Steinbach, M.; Botta, C.; Deubel, T.: Modelling of Product-Service Systems (PSS) Based on the PDD Approach. In: Proceedings of Design 2004, University of Zagreb, 2004, pp. 547-554
[4] Albers, A.; Birkhofer, H. (eds.), „Heiligenberger Manifest, Vorträge, Workshop-Ergebnisse". In: Proceedings of the Workshop „Die Zukunft der Maschinenelemente-Lehre", Universität Karlsruhe / TU Darmstadt, 1997
[5] Pahl, G.; Beitz, W.: Engineering Design. Springer, 1988
[6] VDI-Guideline 2221 (English Version): Systematic Approach to the Design of Technical Systems and Products, VDI, Düsseldorf, 1987
[7] Suh, N.P.: The Principles of Design. Oxford University Press, 1990
[8] Weber, C.: Ein Beitrag zur integralen Betrachtungsweise von methodischem Konstruieren und Maschinenelementen. In: Hubka, V. (ed.): Methodisches Konstruieren der Maschinenelemente, Heurista, Zürich 1986, pp. 78-102
[9] Weber, C.: Neue Aspekte der Analyse und Synthese dynamischer Prozesse auf der Basis der Konstruktionsmethodik – gezeigt am Beispiel „Automatischer Blockierverhinderer für Kraftfahrzeug-Bremssysteme. Konstruktion 39 (1987) 10, pp. 391-400
[10} VDI-Guideline 2206: Entwicklungsmethodik für mechatronische Systeme / Design Methodology for Mechatronic Systems. VDI, Düsseldorf, 2004
[11] Weber, C.; Werner, H.: Schlußfolgerungen für „Design for X" (DfX) aus der Perspektive eines neuen Ansatzes zur Modellierung von Produkten und Produktentwicklungsprozessen. Proceedings of the 12. DfX-Symposium, Universität Erlangen-Nürnberg, 2001, pp. 37-48
[12] Weber, C.; Werner, H.; Deubel, T.: A Different View on PDM and its Future Potentials. In: Proceedings of Design 2002, University of Zagreb. 2002, pp. 101-112
[13] Weber, C.; Deubel, T.: New Theory-Based Concepts for PDM and PLM. In: Proceedings of ICED 03, The Design Society & the Royal Institute of Technology, Stockholm, 2003, pp. 429-430 (executive summary), paper no. 1468 (full paper on CD-ROM)
[14] Burr, H.; Deubel, T.; Vielhaber, M.; Haasis, S.; Weber, C.: IT-Tools for Engineering and Production Planning in an International Automotive Company – Challenges and Concepts. In: Proceedings of the 36th CIRP International Seminar on Manufacturing Systems, Saarland University, Saarbrücken, 2003, pp. 73-80

8 Author

Prof. Dr.-Ing. Christian Weber

Saarland University, Institute of Engineering Design/CAD

PO Box 15 11 50, D – 66041 Saarbrücken, Germany

Phone: +49 / (0)681 / 302 – 3075; Fax: +49 / (0)681 / 302 – 4858

Email: weber@cad.uni-saarland.de

URL: http://www.cad.uni-sb.de

Part IV: *IT Aspects, Methods & Tools*

The Evolution of New PLM Technologies

(Seven Steps from Research to Business Benefit)

George Allen

UGS, Cypress, USA

Abstract: This paper outlines a seven-step model for the evolution of a typical new PLM (product lifecycle management) technology. The basic steps are academic research, incorporation into commercial software packages, deployment within end user companies, and finally (we hope) delivery of business benefits. We illustrate this process with some examples, and examine what might go wrong in the seven steps. Finally, we suggest some ways in which the evolution process might be improved.

1 Introduction

Figure 1 below outlines a simple model of the process through which new PLM technologies ultimately deliver business benefits to end user companies.

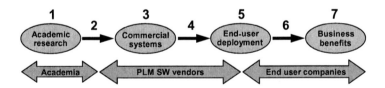

Figure 1: The seven step process

The steps can be briefly summarised as:

(1) **Research**: invention of a new technology, often in academia

(2) **Dissemination**: migration of knowledge into vendor companies

(3) **Adoption**: incorporation into commercial products

(4) **Sale**: products are sold to end user companies

(5) **Deployment**: the new product is deployed within the end-user company

(6) **Operation**: day-to-day usage of the new technology

(7) **Evaluation**: benefits are assessed, publicised, celebrated

In the remainder of this paper, we describe this process in more detail, illustrate it with some examples, and discuss its effectiveness. Ultimately, we would like to improve this process, and learn how to be more successful in deriving business benefits from the technology we develop. A failure at any step means that the entire process is a failure, so we need to be successful at every stage. This can only be accomplished through a holistic consideration of the process by all three of the communities involved: academia, software vendors, and end users.

2 What Can Go Wrong in the Seven Steps

In the seven items in this section, we outline some of the things that can go wrong in each of the seven steps of the process outlined above.

2.1 Research Failure: Academic Research Irrelevant

In most cases, academics (especially in computer science or mathematics departments) have very little involvement with end user companies, so it is difficult for them to understand the real problems of these companies. As a result, academic research is sometimes irrelevant – the problems addressed are not important, or adequate solutions are already available. For some academics, the business relevance of their research is not even important. Their role, as they see it, is basic research, so they are focused on the acquisition and codification of knowledge, not delivery of business benefits. This is certainly a tenable point of view. Some would argue that if research is targeted at providing directly usable results to industry (either the software industry or the manufacturing industry) then more of it should be funded directly by industry.

2.2 Dissemination Failure: Vendors and Academia Disconnected

While this varies between countries and disciplines, academics and commercial software vendors are sometimes not well connected. The disconnect goes in two directions: commercial software developers are not aware of what academics are doing, and vice-versa. Part of the problem results from the way research results are described and published. The academic community (especially in mathematics) places a high value on abstraction, but abstract results are difficult for the average software developer to understand and appreciate. A busy programmer will typically say he is "too busy" to read research papers that might be relevant to his work, especially if the papers are difficult to understand or their relevance is not immediately obvious. To make matters worse, many programmers find it enjoyable to invent their own solutions, rather than adopting approaches developed by others. Inventing is fun, even if it is re-inventing.

Good connections are guaranteed when people move from commercial software companies to academia or vice-versa. The former case is fairly uncommon because salaries in academia are typically far lower than in industry. The latter case (academics moving to industry) works especially well when new research forms the foundation for start-up software companies. In some countries, this is a fairly common occurrence, but it works best if there are government programs or a well-established venture capital community to support it.

2.3 Implementation Failure: Solutions Difficult to Incorporate

Some new technologies simply do not fit well into existing commercial systems. Existing systems are large and complex, and making fundamental changes to their architectures is difficult and expensive. To justify these large changes, the new technology has to deliver enormous benefits, and this is rare, of course. Two examples are technologies that require unusual new geometric representations, or ones that require special hardware devices. These are not likely to find their way into established commercial systems, though they might be adopted by start-up companies who have less inertia.

2.4 Sales Failure: New Ideas Difficult to Sell

Technology, no matter how marvelous, does not sell itself. Sales people must be educated, so that they can explain the technology and its benefits to prospective buyers. If neither the salesman nor the prospect understand the offering, it is not likely to get sold. Unfortunately, radically new ideas are often quite difficult to understand, and are therefore difficult to sell. For this reason, commercial software vendors are often quite conservative about pursuing new technologies – obviously it does not make sense for them to create products that they are not able to sell effectively. Of course, even the most expert salesman will not be able to sell a product effectively if the market for it is small or impoverished.

2.5 Adoption Failure: New Technology Difficult to Adopt

Visionaries in end-user companies might see the benefits of some new technology, and promote its use. But if the technology (or its implementation) can only be understood by far-sighted visionaries, and not by the average employee, then it will not succeed. New approaches sometimes require business process changes, personnel changes, or organizational changes, and this much change is difficult for many companies. Conservatives would argue that software solutions should adapt to existing business processes, not the other way round, but this makes new approaches difficult. High-end PLM vendors can often help with training, implementation plans, and business process re-engineering, but these services are sometimes expensive, and many companies are unwilling to pay for them. In highly stressed industries, management often demands immediate return on their investments – they argue that if improvements are not delivered in the short term, then there will be no long term. But dramatic technological changes are often disruptive and counter-productive, in the short term, which makes them unacceptable in some situations.

2.6 Operational Failure: New Technology Doesn't Deliver Benefits

Sometimes new technologies are simply over-sold, both by vendors and by internal advocates. It is easy for people to become excited by new technologies, and to imagine that they will solve all manner of problems. But the initial implementations of new technologies often do not work very well -- what works in a demo often does not work in real applications. As a result, early adopters frequently experience difficulties and disillusionment. Even if the new technology does work, it may not deliver any significant business benefits. The visionaries who investigate and promote new technologies sometimes do not understand their company's business well enough, and may spend lots of time and money implementing solutions to problems that have little impact on business success.

2.7 Evaluation Failure: Business Benefits Unappreciated

A new technology can only be declared a success if there is some reasonable way to assess the business benefits it provides. In many companies, there are no metrics in place to measure the performance of critical processes, so improvement is difficult to measure, too. Some of the benefits related to PLM are very intangible and difficult to measure -- how can we quantify "better products" or "increased innovation", for example? Ironically, some of the best successes are often kept secret. Managers fear that if they announce a 30% improvement in productivity, they will be asked to cut their budgets by 30%. Anyone who deals with organized labor (unions) has to be especially careful about discussing new technologies or improved productivity, since these are often equated with loss of jobs.

3 Example 1: Basic Geometric Modeling

3.1 Current Situation

Currently, almost all commercial systems use fairly conventional boundary representation modeling, using data structures like those shown in figure 2 below:

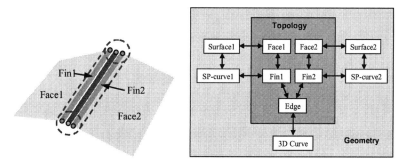

Figure 2: Basic data structures

Various body types are supported, including solids, sheets, and more general bodies such as non-manifold ones [9]. The basic geometry types typically include quadrics, tori and extruded and revolved surfaces. In addition, most systems support some fairly general form of tensor-product non-uniform rational b-spline (NURB) surface. In some systems, there are special procedural geometry types, such as procedural offset surfaces and procedural intersection curves. Some support for "foreign" geometry (user-supplied evaluators) is also available in some systems.

3.2 Progress Through the Seven Steps

3.2.1 New Geometry Types

New geometry types are frequently studied in academia. These include triangular patches, subdivision surfaces, cyclides, trigonometric and exponential splines, and so on. But the existing geometry types listed in the previous section seem adequate already, and new types bring few additional benefits. Surfaces with non-rectangular parameter domains (such as subdivision surfaces and triangular patches) are extremely difficult to implement in main-stream commercial systems, so the return on investment is correspondingly low. So, here, we have failures in the research, dissemination, and adoption steps.

3.2.2 Unconventional Shape Representation Schemes

New primary shape representation schemes arise from time to time. Some examples are SGDL [7], voxel/octree representations [4], etc. These have only minor advantages (if any) over conventional boundary representations, and they are difficult to add to existing systems, so they are not likely to be adopted. In particular, there is still considerable research into divide-and-conquer algorithms based on traditional CSG (constructive solid geometry) representations [1]. This work is largely irrelevant. Commercial systems do not use CSG-based divide-and-conquer algorithms, and are not likely to. The failures here are in the research and dissemination steps.

3.2.3 Tolerance Control

Control of tolerances in geometric calculations is an important issue that receives quite a lot of attention in the research community. Having been refined by millions of hours of production use, the approaches used in commercial systems work reasonably well in practice, but they are generally without rigor or a scientific foundation. Much of the research work focuses on the use of exact rational arithmetic or interval arithmetic [2]. While this might provide important theoretical insights, these techniques are far too slow for production usage, so we have primarily dissemination and adoption failures.

The Evolution of New PLM Technologies

4 Example 2: Parametrics and Associativity

4.1 Current Situation

The basic ideas of parametrics and associativity are well known. Briefly, each object retains a "recipe" -- a record of its inputs (parents) and the operation that was used to create it. This allows automated reconstruction of the object following changes to inputs, thus preserving "design intent". Associative modeling operations may be linked together in sequences, with the output of one operation being used as input to next, so that changes can be propagated from parent to child, to grandchild, and so on. By capturing sequences of operations in a replayable form, we preserve valuable design process knowledge. The ideas are illustrated in figure 3 below:

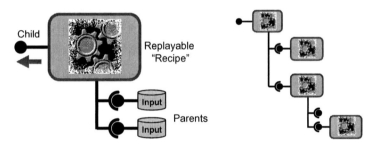

Figure 3: Associativity

The ability to replay a sequence of operations with new inputs allows us to automate the rework that is required when design changes are received from upstream. For example, when the outer surfaces of an automobile hood are changed, we can "replay" the construction of the hood inner panel. We may even be able to generate a hood inner for a new vehicle by replaying the construction of the hood inner from a previous model. Or, instead of just re-using previous designs, we may construct "template" models that are specifically designed for re-use. This is not an easy task, but the benefits are very large, so this approach has been pursued in a number of different industries. For example:

(1) General Motors has produced template models of vehicle architectures

(2) General Electric has template models of jet engines

(3) Boeing has template models of wing structures

(4) A company called The Alloy has built template models of cell phone interiors

GM – vehicle architectures (1994)

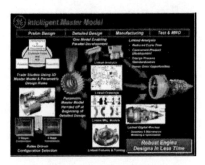
GE – jet engines (2001)

Boeing – wing structures (1998)

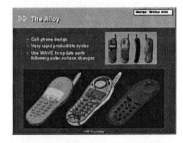
The Alloy – mobile phones (1998)

Figure 4: Examples of template parts

While parametrics, associativity, and template parts are being used successfully by many companies, the technology is not without problems. Transferring parametric models between different vendor's systems is impossible, except for very simple designs. Also, parametric models sometimes fail to replay correctly, and identifying the causes of failures is usually quite difficult, especially if the model was originally built by someone else.

4.2 Progress Through the Seven Steps

4.2.1 Scientific Foundations

Parametrics and associativity were developed mostly in industry, not in academia. As a result, the scientific foundations of the technology are weak. Vendors have typically developed ad hoc solutions to problems, and ignored academic efforts [5]. Commercial systems incorporate very few formal definitions of concepts or of "correct" behaviour. So, in this area, there are failures in the research and dissemination steps.

4.2.2 Data Exchange

Data exchange between two parametric systems is a very difficult task [6]. The semantics of an associative object are usually embodied solely in the code used to generate it (and regenerate it). Therefore, reproducing object semantics across systems requires sharing of code, and there is no reasonable way for competing vendors to do this. Data exchange problems deter implementations of the technology in end-user companies. The primary failures here are in the deployment and operation steps.

4.2.3 User Expertise

Designing a re-usable parametric model is a difficult task -- certainly much more difficult than designing a static single-use model. Real-life examples are complex, and never work as well as the simple vendor demos. Even with the best training, some users are simply incapable of learning and applying the technology effectively. In some companies, the solution is to form an elite group of specialists who develop template parts for use by other less capable users. But this requires organizational change, which is difficult in some companies. So we typically find problems in the deployment and operation steps.

5 Example 3: Class A Surfacing

5.1 Class A Surfacing – Current Situation

In the latter part of the automotive styling process, specialists develop "class A" surfaces. These are the final production-quality versions of the aesthetically significant portions of the vehicle (exterior surfaces, mainly). These surfaces are then used in the development of structural parts and tooling. In most companies, software used for this purpose is somewhat fragmented – different systems are often used for conceptual design, class A surfacing, and downstream engineering. Systems used for class A surfacing typically have a wide variety of curve and surface editing functions, plus real-time functions for evaluating the quality (curvature, smoothness) of shapes, as illustrated in the figure below:

Figure 5: Class A surfacing

There is a great deal of academic research focusing on the mathematical and computer graphics techniques involved [3]. In most companies, surface development uses a combination of physical (clay) models and digital models, but physical models are almost always used for final design approval.

5.2 Progress Through the Seven Steps

5.2.1 Geometry Construction Functions

Despite the large body of research on geometry construction and modification techniques, most of the real work is done using very simple functions. In fact, simple manual movement of Bezier or b-spline control vertices is by far the most common technique. There seems to be a failure here in either research or dissemination.

5.2.2 User Skills

Learning to use class A surfacing software is not easy. Moreover, the people who have traditionally done this work are typically craftsmen who have limited familiarity with computers, making the learning process even harder. So, in many companies, there are plenty of people who know how to manually modify a clay model, and very few who know how to modify a digital one. Faced with this situation, many companies have been slow to adopt digitally-based processes – there have been failures in the deployment and operation steps.

5.2.3 Business Viability

Several vendors of class A surfacing technology (Cisigraph, CDRS, Imageware) have gone out of business or have been absorbed by larger main-stream CAD companies. The fundamental problem is that the market size for class A surfacing software is not large enough to support the development investment required. The problem has become worse in recent years now that PC-based systems are consuming the low end of the market. So, for a software company, developing class A surfacing technology is not a very attractive business. This can be viewed as a failure in the sales step of the process.

5.2.4 Surface Evaluation Functions

As indicated above, Class A surfacing packages typically have very sophisticated functions for assessing the "quality" of surfaces. Typically, this means that the functions can detect various types of shape discontinuities, usually through the use of simulated reflections [8], as shown in the figure below:

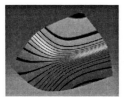 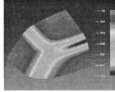 

Figure 6: Surface evaluation functions

In some ways, the use of these functions is counterproductive. The functions are far more sensitive than the naked eye of even the most discerning car buyer, so users spend long hours correcting "problems" that are probably invisible in the show-room. On the other hand, there are surface attributes such as overall flow and proportion that can not be evaluated by software functions, and which probably have a greater impact on the marketability of vehicle. There is little evidence to indicate that surface continuity is a significant factor in automobile sales, so we have a failure in the evaluation step of our process.

6 How to Improve the Situation

This section briefly outlines some suggestions for reducing the incidence of failures in our seven-step evolution process.

6.1 Increased Interaction

It would help if there were more interaction between researchers, vendors and end users. For example, academics should gain some experience using commercial systems, or at least some awareness of their capabilities and limitations. Software vendors are often willing to give their products to universities, so access should be inexpensive. Academics in departments of engineering, business, and economics often have better connections with industry than those in mathematics and computer science departments. Therefore, increased interaction between different university departments would be valuable. On the other hand, software vendors should spend more time studying research literature, and understanding its relevance to their problems and their customers' problems.

6.2 PLM Software Vendors as Process Consultants

PLM software vendors must continue to be process consultants. They must have a clear understanding of their customer's businesses, and must be able to guide them through implementations that will deliver business benefits. PLM offerings need to be a combination of software and services. Of course, the services have to be affordable, or they will not be used. It is ironic that companies who are willing to spend money on software are not willing to invest in expert deployment help. Without a good deployment strategy, the money spent on software is largely wasted.

6.3 Simpler Software

Software systems need to be simpler and easier to deploy. A great deal of effort is expended on improving user interfaces, but much of this effort focuses on making the user interface prettier, rather than improving its conceptual simplicity. System functions are generally much easier to understand and implement if they are task-specific, rather than generic. So, for example, doing mold design with a specialised mold design package is typically much easier than using a generic geometric modeling package. Task-specific functions can use terminology that is more familiar to the user, can use appropriate default values, and can be more highly automated. All of these make the software easier to understand and deploy.

6.4 Acceptance of Change

While software vendors should make every effort to develop solutions that fit into existing working environments, there are times when this is not possible. Sometimes, the full benefits of a new technology can only be realised by making significant changes in processes or even organisational structure. This also implies that the benefits of the technology will not be immediate, so some patience may be required.

7 Conclusions

As we have seen, the seven-step journey from academic research to end-user business benefits is along and tortuous one. There are many places where failures can (and do) occur, and one failure means that the entire process is a failure. Success requires effort from the research community, commercial software vendors, and the end user companies themselves. While this paper has focussed on possible failures, there have also been many successes. In general, the keys to success seem to be communication across the three communities mentioned above, and the breadth of understanding that this communication promotes.

8 References

[1] CSG 98: Set-Theoretic Solid Modelling, Techniques and Applications, Proceedings of the CSG 98 Conference, Ammerdown, UK, April 1998. ISBN 1-874728-12-7

[2] Christoph M. Hoffmann, "The Problems of Accuracy and Robustness in Geometric Computation". Computer 22(3), March 1989, pp. 31-41.

[3] Josef Hoschek and Dieter Lasser, "Fundamentals of Computer Aided Geometric Design". A K Peters, 1993.

[4] C. L. Jackins and S. L. Tanimoto, "Oct-trees and their use in representing 3D objects", Computer Graphics Image Processing 14 (3) (1980), pp. 249-270

[5] Srinivas Raghothama and Vadim Shapiro, "Topological Framework for Part Families," Transactions of ASME, Journal of Computing and Information Science in Engineering, Volume 2 (2002), No. 4, pp. 246–255.

[6] Steven Spitz and Ari Rappoport, "Integrated feature-based and geometric CAD data exchange". Proceedings, Solid Modeling '04, June 2004, Genova, Italy, ACM Press.

[7] Jean-François Rotgé, "L'arithmétique des formes: une introduction à la logique de l'espace". PhD thesis, Faculté d'aménagement, Université de Montréal, 1997.

[8] Holger Theisel and Gerald Farin, "The curvature of characteristic curves on surfaces". IEEE CG&A, Vol. 17, No. 6, Nov. 1997, pp. 88-96.

[9] Kevin Weiler, "Edge-based Data Structures for Solid Modeling in Curved-Surface Environments", IEEE CG&A, Vol. 5, No. 1, Jan. 1985, pp. 21-40.

9 Author

George Allen, UGS, 10824 Hope Street, Cypress, CA 90630, USA.

george.allen@ugs.com

Product Development System with Special Focus on Automotive

David Blair

PTC, Milano, Italy

Abstract: As tools for product development have evolved, it has become clear that non-integrated point solutions are falling short in delivering ROI. As these types of systems are implemented, engineers must adapt to various user interfaces for these tools, administrators must maintain various systems on different architectures, and people interested in seeing a business process completed from end to end are not always sure it can be accomplished. Considering these factors, while innovating new approaches to product development, PTC has created what it calls a Product Development System (PDS). What are the components of this system? Why does this system bring value to customers? What can we expect from product development tools in the future? This discussion will provide insight on how PTC is creating a product development system that is ready to bring new value today. And for those with a mind towards automotive, you will also hear how PTC is developing it's system to supply automotive OEM's, like Toyota, with competitive advantage far into the future.

1 Introduction

This paper will cover the latest strategy and technology offerings from PTC. PTC is a worldwide software vendor of Product Development tools with over 35,000 customers around the world and is focused on creating a Product Development System applications for creation, collaboration, and control in a single unified software architecture. The topics covered in this document include the following:

- PTC Vision
  - Product Development System (PDS)
  - Benefits of Taking a System Approach
- PTC Technology
  - Create, Collaborate, Control
  - Bringing it All Together
- PTC in Automotive
  - Next Generation System
  - Advanced Features

2 PTC Vision

2.1 Today's Product Development System

Today's product development environment can look very chaotic and unorganized because many times these systems grew up as piece parts over time. In the figure below the lines represent varying degrees of communication and different types of communication in a typical product development environment. There is product information all over the enterprise and outside of the enterprise. It is very difficult to implement new strategies and initiatives in this environment. However, on the right side of the figure below, one sees that PTC's has vision for how a product development system should be - a single integrated system with all digital product information managed and controlled.

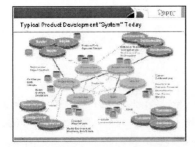 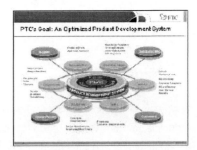

Figure 1: Today's typical product development environment compared to a system approach being developed by PTC

2.2 Critical Factors for Success

PTC's Product Development System (PDS) is product development software planned, developed, and tested as a system containing the right capabilities and architecture enabling seamless completion of business processes end-to-end. Ultimately, this system is delivered with a proven recommendation on how to implement the technology.

PTC believes that a successful Product Development System must contain the right capabilities including creation tools (CAD/CAM/CAE), collaboration tools for share data with suppliers, customers, development partners, etc., and control tools to manage all of the data in a well organized data management environment. In addition to having the right capabilities, the PDS system must also be built on the right architecture. The architecture should be integral (single database), internet based (easy to deploy and takes advantage of standard protocols, and interoperable (able to easily integrate with other systems like ERP, etc.

2.3 Benefits of a System for Product Development

A Product Development System that is architected, developed, and tested as a system will provide a better return on investment because it offers value with it capabilities, lowers investment with well defined implementation packages supporting it, and lowers risk by ensure the system works together. In addition to these things, by having a "system" that works together on the right architecture, it will lower administration and integration with other systems costs.

It will also improve user adoption because a web-based paradigm is well understood and a system with similar GUI's will be easier to learn.

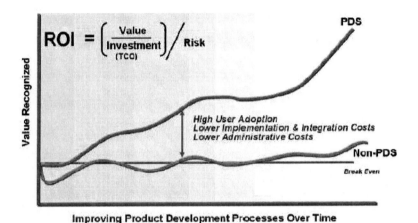

Figure 2: A system approach to Product Development Tools offers greater ROI

3 PTC Technology

3.1 Create, Collaborate, Control – Why You Need All 3

PTC has worked to build out a strong PDS platform that provides for creation, colloboration, and control tools for product development. Before expanding on what these solutions do, it is important to understand why it is critical to have all three of these capabilities. Creation tools like Pro/ENGINEER for CAD/CAM/CAE are important because they allow for creation and maximizing innovation (more exploration of ideas). Collaboration tools like Windchill ProjectLink allow for design collaboration with customer, suppliers, partners, etc. Better collaboration enables more feedback and the ability to shorten design cycles with better communication. Finally, it is important to have control tools like Windchilll PDMLink to vault, manage, version control, etc design data in order to drive convergence of a product design for release to manufacture.

Product Development System with Special Focus on Automotive

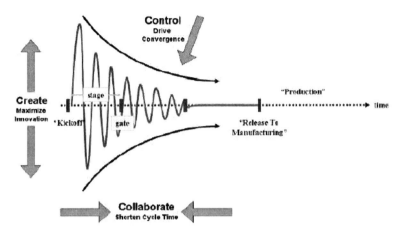

Figure 3: Why you need creation, collaboration, and control

3.2 PTC Solutions

The PTC solutions for create, collaborate, and control are Pro/ENGINEER Wildfire, Windchill ProjectLink, and Windchill PDMLink. The capabilities for these tools that make up the PDS from PTC are as follows:

Pro/E Wildfire Capabilities (CREATE):
- Simple
    - New user model with 20-80% productivity improvements
- Powerful
    - All of the rich capabilities needed for design of products
- Connected
    - Worlds first multi-user CAD system
    - World's first intelligent, on-line, process-oriented CAD system

Windchill ProjectLink Capabilities (COLLABORATE):
- Web-based project portal solution with embedded 3D visualization, search
    - Self-administration project management
    - Access anywhere to project data through intuitive Web browser-based user interface

- Product development process control
    - Process templates: design reviews, NPI, quality (Six Sigma, APQP, ISO 9000), etc.
    - Connects to various CAD systems as well as Microsoft Project
- Configurable Quick Start implementation

Windchill PDMLink Capabilities (CONTROL):
- Product information control
    - Global access to all product data including designs, models, specifications, and requirements
    - Personalized information and product data
    - Visual navigation of product structures to easily find information
- Product development process control
    - Robust workflow engine enables repeatable and predictable product development processes
    - "Out-of-the-box" CMII change control process
    - Visually-driven configuration management
- Enterprise-Wide access to product information
    - Access anywhere to product data through intuitive Web browser-based user interface
    - Connects to various CAD, MRP, and ERP systems (CAD, ECAD, Software) – all aspects of product definition
- Configurable Quick Start implementation

3.3 Single Application View with PTC PDS

By developing these applications above with a view to building them as a single Product Development System (PDS), PTC has been able to create a single application view to getting at all of this powerful, yet easy to use software. From the CAD environment, a user can get to CAD/CAM/CAE tools, Windchill ProjectLink for project collaboration and management, and Windchill PDMLink for PDM needs.

Product Development System with Special Focus on Automotive

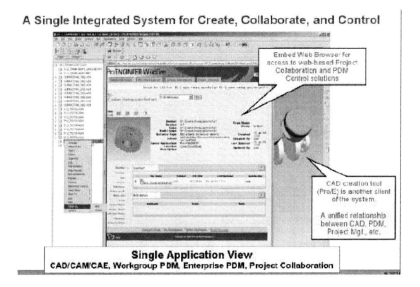

Figure 4: Bringing it All Together – PTC Delivering the PDS

4 PTC in Automotive

4.1 Presence in Automotive and The Toyota Project

PTC has approximately 15% of the automotive industry as users of its software, accounting for roughly 13% of its revenue. There is a strong predominance of PTC solutions being used in powertrain development. With a strong knowledge of powertrain requirements and a proven track record in this industry, Toyota Motor Corp. selected PTC to be its long term partner in the generation of new CAD/CAM/CAE David Blair software to be used to help it reach its goal of reducing engine development from 24 months down to 12 months

4.2 Toyota Requirements

Today Toyota's requirements are focused on the following:
- Improved Planning / Concept
    - Real time cooperation w/ vehicle team
    - Database / knowledge driven
    - Rapid conceptual design

- Rapid Product & Process Design
    - Real time assembly & component design
    - Virtual prototyping
    - Simultaneous process design
    - Simultaneous tool, jig, fixture design
- Rapid Manufacturing of Tools, Jigs, Fixtures
    - Shorten times from design freeze to start of production
- Collaboration and Control of Engineering Information
    - Product, process, and production

4.3 Next Generation CAD/CAM/CAE for Automotive

Based on Toyota requirements, PTC is developing Pro/ENGINEER Wildfire to add advanced capabilities to speed up engine development. These requirements have been translated into a next generation system with new capabilities such as the following:

- Web-based Database Centralizing and Controlling Design Data and Processes
- Managed Storage and Reuse of Existing Knowledge
- Integrated Quick CAE
- Improved Internal and Supply Chain Collaboration
- Integrated and Associative NC Planning and Machining
- Product Development System with Special Focus on Automotive
- Improved Drawing Capabilities
- Full 3D Annotation to Remove Drawings from Critical Path

5 Author

David Blair, PTC, Palazzo Sirio 3, 20041 Agrate Brianza Milano, Italy
dblair@ptc.com

Software Systems Development and Deployment - Some Aspects out of Daily Work

Anton M. Cremers [1], Peter Reindl [2]

1) BMW AG, München (Germany)
2) BMW AG, München (Germany)

Abstract: This paper focuses on some aspects of software systems development and its deployment in the automotive domain. It identifies how IT effectiveness is driven by corporate strategy and the requirements arising from this strategy. It recommends the deployment of standard software (Commercial Off-The-Shelf software - COTS) instead of custom-developed software. Furthermore, it describes how flexibility into the processes is achieved through the use of standardized interfaces this being the key pre-requisite of process innovations. As software deployment for a vast number of end-users requires a well-coordinated software release process, a generic process with IT Infrastructure Library (ITIL®) is presented. This paper addresses how to avoid risky unplanned events during software development and deployment. To reduce the unacceptable costs on the end-users' side, the use of appropriate quality assurance and release management processes for COTS are also presented.

1 Corporate Strategy Drives IT Effectiveness

IT contributes to the "key success factors" as well as to the "factors of production" of a company. In this respect, IT requirements are derived from Corporate Strategy aims (Fig. 1).

However, current IT systems are often large, monolithic, proprietary islands of software. In response to changing business conditions, the problems associated with this are:

- Limited flexibility (hindering process innovation)
- High development, enhancement and maintenance costs

- High integration costs with other systems (lack of interchangeability and interoperability)

In order to manage these problems, following requirements arise:

1. Standard software instead of custom-developed software should be used. On the other hand, the standard software should be flexible enough to allow innovation of processes and it should be fully integrated to be able to establish mature business process.
2. This can be put into practice with reasonable efforts and costs only with the help of standardized interfaces.
3. During the software life cycle, the Release Management Process should be synchronised with time and with minimum of effort and costs, but without compromising the overall software quality.

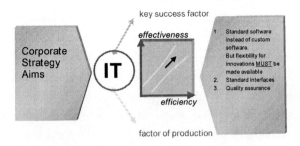

Figure 1: Corporate Strategy drives IT Effectiveness

2 Standard vs. Custom-Written Software

The enterprise IT systems are usually installed by using Commercial Off-The-Shelf software (COTS). These are commercially available products that can be purchased and integrated with little or no customization, thus facilitating company's infrastructure expansion and reducing costs. COTS applications are tempting because they provide turnkey functionality. However, there are several kinds of COTS software, all of which might be applicable:

- Standalone applications
- Libraries

- Application-specific languages
- Development environments with runtime modules
- Device drivers

Clearly, not all of these are turnkey, off-the-shelf tools. Particularly with regard to large PLM systems, some elements of „development" is still needed, except in the smallest applications, as they are never used "out-of-the-box".

Table 1 gives a short comparison between standard and custom-written software.

Standard Software	Custom-Developed Software
Higher initial costs	Lower initial costs
Lower deployment costs	Increased development costs
Faster project implementation	Slower project implementation
Increased reliability	High reliability requires increased in-house development
Standard data exchange	May not fully support industrial standards
Decreased training requirements	Requires extensive knowledge of proprietary systems functions

Table 1: Comparison of Standard and Custom-Developed Software

Due to the perceived difficulties of custom-developed software, the desire exists to have "no custom code" in the applications. In practice, this translates to the requirement to have as much as possible of standard software and as little as possible of custom software in the system.

Despite of its exclusiveness, standard software has to allow:

- Process innovations
- User interface integration possible
- Retention of flexibility in system decisions
- Co-operation with application partners

To satisfy this, standardized system interfaces across applications, systems and providers are needed. This goal, however, is not achieved today.

3 The Role of Standardized Interfaces

The specific processes within a company may imply a significant competitive edge. The use of company's IT will be the more successful the better it exactly meets the process' demands. Thus, the role of well-defined interfaces within the processes becomes crucial, particularly with regard to the flexibility and the speed of process innovations.

For rapid implementation of process innovations, both interoperability between software packages and the flexibility of the "best-in-class" component choice are critical.

As the end-users need to exchange complex information on a mixture of strategic and mundane requirements, they demand more interoperability between applications of different vendors. However, the integration should be possible within acceptable efforts and costs and should be implemented by software vendors or 3rd party software suppliers using (adopting) existing systems and components or by introducing new standalone applications (COTS, see Chapter 2).

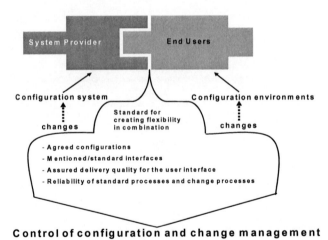

Figure 2: Standards create flexibility for innovation

On the other hand, end-users want to stay flexible in the choice of system components in order to achieve the process innovations within the shortest time. The speed of the deployment of process innovations is an important success factor of the company.

Last but not least, standardized interfaces between different software-packages have to allow both changes to the systems and environment configuration (Fig. 2).

Currently, many software interfaces do not satisfy these demands and limit or hinder rapid innovations and implementation.

4 Generic Release Process

So far asserted that standard software (COTS) compared to custom-developed software has clear benefits with regard to rapid implementation of process innovations. Considering the involvement of vast number of end-users (i.e. 300 or more) a well coordinated software release process is required. Furthermore, it is vital to ensure that software vendors are also fully integrated into this release process to monitor software release quality and other issues related to successful rollout of the software. The way to arrange a professional, well-defined and secure software release process is the use of the IT Infrastructure Library (ITIL®)[1].

4.1 The IT Infrastructure Library (ITIL®)

ITIL is a set of guidelines developed by the Office of Government Commerce (OGC) for the British government. Today, ITIL is the de-facto standard in the area of service management. It contains comprehensive publicly accessible specialist documentation on the planning, provision and support of IT services. ITIL provides the basis for improvement of the use and effect of an operationally deployed IT infrastructure. IT service organizations, employees from computing centers, suppliers, specialist consultants and trainers took part in the development of ITIL®. ITIL describes the architecture for establishing and operating IT service management. The ITIL books are best practice guidelines for service management, with the guidelines describing what rather than how. Service management is tailored to the size, the internal culture and, above all, the requirements of the company. The impartial view of the external consultant may help to break away from the rigid structures.

Release management ensures successful planning and control of hardware and software installations. The focus is on protection of the productive environment and its services by using formal procedures and checks.

The term "release" refers to one or several authorized IT services change measures. Dependencies between a particular software version and the hardware required to run it determine bundling of software and hardware changes which, together with the other functional requirements, constitute a new release.

Release management is carried out based on the Configuration Management Database (CMBD) so as to ensure that the IT infrastructure is up to date. Two further terms defined by ITIL are needed to understand the ITIL release process:

- **Definitive Hardware Store (DHS)**: Tested and released hardware is stored in a secure environment away from the production environment in one or several places. These replacement components and assemblies are kept at the same release level as the systems in the production environment.

- **Definitive Software Library (DSL)**: The content of software releases are documented, and the correctly approved software versions are stored as an autonomous component of the CMBD in a safe environment in the DSL. Backup copies are always stored away from the premises. Furthermore, the DSL contains documentation to identify the release status of all software in use.

4.2 The ITIL Release Process

An ITIL release process consists of following tasks:

- Determine the release policy
- Define and implement releases
- Carry out release tests and acceptance
- Plan and implement rollout
- Provide information and training prior to delivery
- Install new or modified hardware
- Store the released hardware in the Definitive Hardware Store (DHS)
- Release, distribute and install the software
- Store the released software in the Definitive Software Library (DSL)

---

[1] ITIL is a registered trade mark of the Office of Government Commerce (OGC) of the British Government

Depending on the organization and their specific processes, an ITIL-based generic release process can be established (Fig. 3).

The release policy defines the level of the IT infrastructure that is to be controlled by definable releases, the agreed roles and responsibilities, the nomenclature and the numbering system as well as provides delimitation between major and minor releases. In addition, the provisions for an emergency release and a rollback procedure are established.

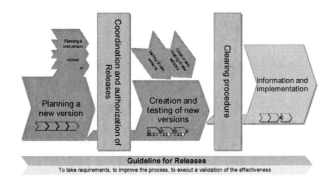

Figure 3: ITIL-based Generic Release Process

Release planning comprises following:

- Define the release content
- Coordinate roles and responsibilities
- Coordinate phases (times, business sectors and customers)
- Prepare a release schedule
- Plan the use of resources (including overtime)
- Prepare backup plan (roll-back)

Furthermore, the release planning shall provide a consideration of the Quality Assurance Process and a quality plan for the release.

5 Software Quality Assurance Process

Quality Assurance (QA) ensures that the developed or deployed software conforms to prescribed technical requirements. QA is essentially an audit process and does not prescribe good or complete testing. It merely documents to what degree developers have followed accepted standardized processes.

The Quality Assurance Process ranges from software development (software vendor) through the software release process until the Start of Production (SOP) and beyond (Fig. 4). It consists of following steps/actions:

1. Ensuring the basic quality of the software before releasing it to the end-users. This should be done by software provider.
2. Software release process directed by the end-user. It includes quality assurance, problem management etc. and alternatively.
3. Actions to cope with unexpected delays in order to keep the SOP.
4. Workarounds as well as compensation actions in case SOP cannot be kept (contractual penalty).

Obviously, this describes a generic process, which needs specific adaptation.

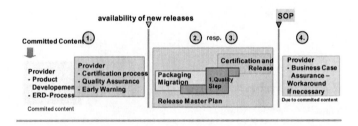

Figure 4: Quality Assurance Process (ERD Evolutionary Rapid Development)

Currently, agreements done ahead of relevant changes are missing in the following process. Thereby, the demands to the tested software quality and its proof are relevant.

The management of (hazardous) incidents during the development, customization or modification of each software component is essential. Otherwise the release changes management of the whole system landscape runs the risk to become incontrollable. This would cause excessive, unacceptable efforts and costs cumbering the end-users.

In case of unexpected problems with COTS, remedies and compensation requests at software vendors can be more often taken into consideration.

6 Conclusions

It is a challenge of Release Management to carry out the software release process within well co-ordinated time and by keeping both costs and quality under control. A well-defined release process, representing all commitments for the management of releases, can be established by means of the IT Infrastructure Library (ITIL). To keep the overall quality of the deployed software on a high level, a quality assurance process has to be defined to conduct the software release process (Fig. 5).

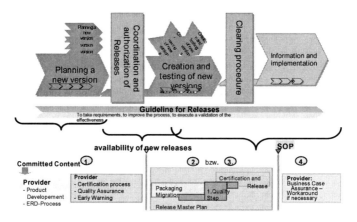

Figure 5: Release Process conducted by Quality Assurance Process

In the release process mainly standard software (COTS) rather then custom-developed software for special applications should be deployed. COTS have the advantage of short-term availability, lower costs and maturity. On the other hand, it should be flexible enough to allow process innovations. This can be put into practice with reasonable efforts and costs only with the help of standardized system interfaces.

7 Authors

Anton M. Cremers, BMW AG, Knorrstraße 147, 80788 München, Germany, anton.cremers@bmw.de

Peter Reindl, BMW AG, Knorrstraße 147, 80788 München, Germany, peter.reindl@bmw.de

Any Future for Parametric Design ?

Alain Massabo

think3, Maria Rosenberg, Germany

Abstract: While Product Life Cycle Management (PLM) importance is growing, some of CAx tools on which PLM should rely, have still limitations. This is the case for the so-called "Parametric Design (PD)". There are some domains, where Parametric Design, as it is, cannot be applied. Moreover it has a set of intrinsic issues that lead to the question: "Is there any future for Parametric Design?" The short following paper aims at presenting briefly these issues and possible workarounds before an attempt for answering the above question.

1 Introduction

It is in fashion to speak about the Product Life Management (PLM) as if CAx tools of other domains on which PLM should rely, have no longer unsolved problems. In particular one of the main tools, that would magnify a PLM interest, is the so-called "Parametric Design (PD)" supposed to record, how CAD models are built and able to be rebuilt from any change of the data parameter of the design.

This paradigm seems elegant and undisputable, thus enforcing the idea of a perfect CAD – PLM world; however the user's daily reality is a little bit different. There are some domains, where Parametric Design, as it is, cannot be applied. Moreover it has a set of intrinsic issues that lead to the question: "Is there any future for Parametric Design?"

The short following paper aims at presenting briefly these issues and possible workarounds before an attempt for answering the above question.

2 The issues

One of the main issues of Parametric Design is due to its definition i.e. the recording of how a model is built and its replay / rebuild when some of its parameters change. By parameter we understand data "pre-existing" to the model design or that are not issued from intermediate design steps. This approach is very interesting to define family of models (up to some limits that we will see below) but forbids its use when modifications are unpredictable.

This is the case in the early stage of style design. The style is almost never made straightforward. Style is a social construction, thus based on several refinements that intend to convey stylists' emotions, company corporate identifiers and/or shapes characters etc. Very often the targeted style change with the time (fashion). It can be observed that even if the initial design were recorded, the downstream requested modifications rarely correspond to a parameter change. For instance highlights in given directions or sections can be subject to change and this kind of data do not correspond with parameters, since in the current state of the art they are only extracted from already existing design models. As a consequence the stylists almost never modify as they have designed thus the interest of Parametric Design for them is disputable.

As a matter of fact to get an effective use of Parametric Design designers should be some kind of programmers. Foreseeing the various possibilities of modifications – tuning of the model, they should prepare/build models to accept some anticipated evolutions. We are far from the emotional creativity …

To illustrate the above let us take the examples of the mouse and the car below. How the designer could have built his model to make it possible the different mouse attitudes or car shapes as shown?

Any Future for Parametric Design ?

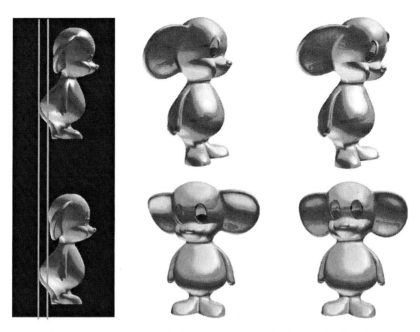

On the left picture the upper part of the mouse (from the belly to the head) is a little bit bent forward while on the four pictures only the head is bent and/or twisted

The nose of the car is becoming sharper from left to right

The back of the car is enlarged on the right exaggerating the shape character

There are some possibilities to modify the history playing with the recorded command graph but unfortunately the range of history changes is quite small.
Consequently Parametric Design introduces a rigidity in the downstream treatments that limits designer's freedom and constrains his creativity.

In addition to the above mentioned drawbacks Parametric Design suffers of:
- A lack of command input understanding / recognition (called tagging problem).
- Model exchanges.
- Model Upward – Downward compatibility.

The first two points are visible by users, while the last one is rather "reserved" to the system developers, even if the users might see the effect on long-term.

The tagging problem is due to the fact that when recording the way the design is performed, the systems build a graph that (in short) put in relations inputs, commands (algorithms) and outputs that at their turns may become inputs for further commands. The connections rely on users' graphic selections i.e. users' choice that the system must understand (guess), in order to be able to replay them, when a rebuild is required after parameter changes. Here are problems, because when making the guess of the users' choice the system must identify the chosen data (the tagging mechanism) such that it can be recognized during the next replay. Unfortunately the users' choices are often context sensitive hence the guess process is not really decidable! Especially when the cardinality of a given commands might vary largely and the user has chosen a sub-set of its output as input for another command. Note that this happens often in complex free form design i.e. in the early stages of the style design. The result of this problem is that the users do not always get what they expect when rebuilding a parametric model. In the best case they can repair locally and in the worse they must redesign the model in a different way. This is "not very good" for the artistic – aesthetic creativity!

Regarding the model exchanges between CAx systems, of course, one can argue that staying in the hands of a single system provider would suppress this drawback but unfortunately would induce some others. Anyway, if we consider that exchanges between systems are useful, then we must recognize that the higher the semantic to exchange the lower the probability of success. Currently, i.e. after more than 30 years of CAD, there are still issues in exchanging pure geometry (despite what the marketing of CAD vendors let us suppose), it is a bit more complicated for the so-called "topology" exchanges and consequently much more for parametric models, in which a large part of semantic is embedded. Unless all systems share the same set of semantics (but then where the differentiation would be?) this point will not be really solved hence it reduces the interest of Parametric Design to its use through a single CAx.

The model compatibility is not so known by the market but is a constraint (sometime a nightmare) for the system developers. Once a Parametric Design model exists it must be usable in all further releases of the system otherwise users loose the fruit of their works. This constrains the evolutions of the functionalities of the system, making its evolution heavier and heavier as the time runs. There are two reasons for this:
- All functionality evolutions cannot be anticipated since the beginning, therefore sometimes significant leaps require to address the upward compatibility issue. In general since the evolution is (fortunately) under developers' control the impact can be reduced.
- There are bugs! They are (unfortunately) not predictable and their fixes constitute real brakes for system evolutions due to aspects of Parametric Design. It cahappen that a bug fix in a new version makes the old Parametric Design model not re-playable or producing a different result. To avoid this, the developers proceed to an algorithm versioning i.e. in short the new version contains the old and the new or updated algorithms and executes each of them according to the input data origin (old or new).

One can imagine that after some times the heaviness of the code that in fine the users suffer in a way or another.

We have seen problems that restrict the interest in Parametric Design and finally make it not so attractive for the users as well as for the developers. Let us see if there is possible way to work around these difficulties.

3 Workarounds

The following techniques present at the first sight advantages against Parametric Design for domains where unpredictable model modifications are a must:
- Variational design, consisting in setting conditions – constraints that the models should fulfill. These conditions are set once or on-the-fly the model designed. Then local modifications are propagated through the conditions to preserve their satisfaction.
- "Tweak" is the possibility for the users to select meaningful topological subsets of the model and to modify/displace them in the model including topological changes (when relevant).
- Global Shape Modeling (GSM), based on 3D morphing defined by constraints that the modified shape must fulfill.

Despite these techniques seem to be satisfying in their concept they have drawbacks as well:
- Variational design is not really applicable on complex free form design where some useful conditions are difficult to express and even more difficult to fulfill. Moreover this technique requires that the set of conditions define a non ambiguous solution. Unfortunately there are under and over constrained situations that are difficult for the users to understand and to correct. Finally being obliged to state the conditions/constraints is bothering and if we expect to guess them then we have to face a problem "equivalent" to the tagging one.
- "Tweak" is limited to a rather small set of operations to avoid a combinatorial explosion of cases. The modification/displacement of the selected subpart is possible but often up to a target "topology" compliancy, for instance the part supposed to receive the displaced subpart must present the same or a compatible "topology" as the one, the subpart has been extracted from. Finally this technique is quite satisfying for mechanical design but not really useful for complex free form design.
- GSM is well suited for complex free form design, gives a lot of freedom to the users (the above illustrations were performed using think3-GSM) but has some numerical limits that bound the number of constraints and unfortunately works only for iso-topology changes.

4 Proposals

We have seen that there are some techniques that help the users when unpredictable modifications are required, each of them having its own domain of application and set of limits. A system that would gather them would reduce their drawbacks and provide some more freedom for creative users.

To secure the Parametric Design interest that PLM let us expect, we suggest the following improvements:

- Since recording how a model is designed is a good and safe way to make the system apprenticeship, it is natural to try to combine Parametric Design with these other techniques. However the first improvement to bring to Parametric Design is to make it optional. The users should be able to decide when and what to record, thus suppressing time to time the Parametric Design rigidity.
- In addition, as nowadays it is the case in some CAD systems, Parametric Design could coexist with Variational Design. Both modes are quite compatible if Variational solving is considered as a node of the history graph.
- Then "Tweak" and GSM techniques should be considered as elements of the history i.e. instead of trying to "program" a Parametric Design model to accept various "unpredictable" modifications, the users could modify as they want after the initial design is performed and record the modifications i.e. keep them in the model.
- Finally the tagging problem could be reduced by giving the users the possibility to explain after or on-the-fly of the design why they select this or that input. For instance the user could indicate that the shortest edge of a face is the spine of the fillet or such selected vertex is the closest to another hole center, etc.

Another (not exclusive) possibility to make Parametric Design able to absorb unpredictable modifications would be to consider the history graph as an input data for Artificial Intelligence technique treatments. This could help to extract "hidden properties or relations" from a given Parametric Design model, to reorder, compare, optimize, substitute subparts of history graph hence making Parametric Design more versatile thus less rigid.

Unless Parametric Design becomes transparent for users especially creative ones, it will stay confined in some application domains thus with a limited future.

5 Author

Alain Massabo, think3, Boulevard Stainte Lucie 39, 13007 Marseille
alain.massabo@think3.com

Long Term Archiving and Retrieval (LOTAR) of Digital Product Data

A.Trautheim [1]; Millarg, L. [2]

1) PROSTEP AG, Darmstadt
2) PROSTEP AG, Darmstadt

Abstract: Long term archiving of CAD data meant so far archiving of 2D drawings as paper drawings, digital files of (usually) tiff format or aperture cards. Since technical products like cars or air-crafts are defined and developed with 3D CAD Systems and Digital Mock-Ups the basics for the technical documentation have changed for certification, legal and customer support. Today 3D models are used as master models. As the result 2D models are not longer fully dimensioned (simplified drawings) and the generation of 2D drawings diminishes in numbers even faster with the aim to get rid of drawings. A retention of the huge number of 3D models in their native formats with write-protected databases or other media is no solution. Because the necessary migration latest after 10 years is too expensive and requires new releases. The project groups "Aerospace LOTAR" (ProSTEP iViP association and AECMA-STAN) and "LZA" (VDA) develop processes and solutions for data preparation, archiving, administration, retrieval and re-use after years of product defining data mainly for the proof of legal, certification and product liability constraints.

1 Long term archiving thousand years ago

About more than three thousand years ago, the Egyptians searched for methods to archive information for later generations and forever. Pharaoh Amenhotep prays to Thoth the god of learning and the patron of scribes and science and asked him: "May Thoth teach him to excel and succeed in the god's ways " (i.e., to be a good scribe). This prayer was written in cursive hieroglyphs on limestone, using a schema of scripts and the material stone as medium suitable for long term archiving (Figure 1, left). It is one of the approaches for "archiving" knowledge for long term.

To certify the authentic of the hieroglyphs, Pharaohs started already to "sign" the stories written on temple walls or in grave chambers of the pyramids. But, the hieroglyphs themselves were changed from period or pharaoh establishing the differences of epochs. Ramses III improved the methods for long term archiving much more. He implemented more than 3.000 signs within the temple of death - beneath than everything else to avoid modification or overwriting for safeguarding the authenticity of his archived "data". (Figure 1, right)

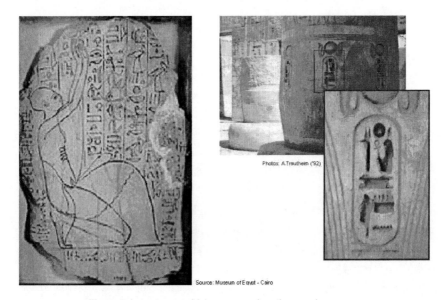

Figure 1: Long term archiving approaches thousand years ago

It can be said, that the Pharaohs developed first methods for long term preservation and the safeguarding of authenticity by using different schemas of hieroglyphs, a purposeful redundancy of data and "technologies" for safeguarding the authenticity of the data. However, they "forgot" to preserve the "Schema", by means the knowledge base to understand the written information after years within a retrieval process. Therefore, it took hundreds of years to decode the different schemas of hieroglyphs and the data content.

2 Requirements and challenges

The requirements of archiving today are much more complex because of the variety of digital, technical product data like native CAD models, measurement, simulation or Product Data Management (PDM) information along the entire life cycle of a product.

The current archiving situation is characterized on the one hand by defined processes and methods for the paper based documentation and on the other hand by new development methods and processes within the digital world. Usually digital archiving processes are based on different CAD/PDM data formats, documents and company dependent structuring mechanisms, which are not standardized and lead into variant preservation problems. These problems are often not existing in the 2D paper based world. For example, the fully measured drawings can be retrieved at any time and additional PDM information are written within the same framework. But these additional information are not available explicitly in a drawing extracted from a 3D CAD model. Therefore the retrieval of attached PDM data, e. g., release status, manufacturing or workflow information are not necessarily part of the drawing, because these data are managed often in a separate PDM system. In addition, the 3D models become more and more the master of the product definition and supersede the 2D drawings. So the changes of the product development tools and documentation methods by using digital data leads into the actual archiving problems (Figure 2).

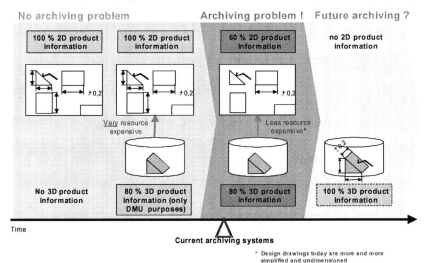

Figure 2: Current and future archiving problems

Currently companies archive their digital data in native formats and databases, because of missing technical, organizational and legal solutions for the archiving of 3D and PDM data. But it is not stable for the necessity of providing the information over years, e.g., in case for a product liability recourse claim.

This results into the requirements, which future digital archives have to fulfill.

These requirements have been harmonized between the European and American aircraft authorities and shall demand all legal and certification relevant documents:

- Ensure continued readability
- Authenticity and identity of the records
- Demonstrate to the authority proper functioning of the records system
- Maintain the capability to retrieve type design data in a usable form for the life of the Type Certificate

Furthermore the following requirements can be identified for long term archiving:

- Regulatory agencies require guaranteed longevity for a variety of digital data objects (type design data)
- The data must be preserved and be accessible in a usable form during the operational life of the product
- Failure to meet this requirement can lead to potential litigation costs associated with being unable to retrieve and use the data
- Preservation and accessibility of design and PDM data is very important for design re-use
- It is critical that the data is retained and accessed with appropriate security mechanisms in place
- It is important to ensure that the retrieved data is in a usable form.

2.1 Technological life cycles

The short-term storage of digital product data in the origin, native data format may fit the former defined requirements, but the constant usage of design data along the entire life cycle results into new aspects which shall be regarded.

At least three technology life cycles need to be considered:
- Product life cycle
  more than 50 years of the products defined by the design data
- Storage life cycle
  last for 10 years of the technologies for storage & retrieval of digital data
- Application life cycle
  approximately 3 years of the technologies used to interpret the data

Therefore to meet the retention requirements, the archived design data have to be resistant against:
- Changes of application technology more than 17 cycles and
- Changes in technology for storage more than 5 cycles

The life cycle of aircraft design data can last up to 100 years. The military B52 bomber is an example for this. The first concepts were written in the 1930ties. The development of the program will be extended until 2025, so that the last aircraft will fly maybe until the middle of the century.

At least the latest developments for the program will be done in 3D. The retention period for this kind of data will be up to 30 years after the last flight.

2.2 Challenges of 3D archiving processes

The storage of native data formats contains a lot of problems and does not fit the requirements for a long term archive. The storage of native formatted data includes the migrations into new formats which could lead into wrong or even no results already within a single application cycle, e. g., when loading in a new generation of CAD-System (3...10 yrs).

The following figure shows an interpretation mistake by loading a CATIA V4.22 CAD model into an CATIA V5R7 environment.

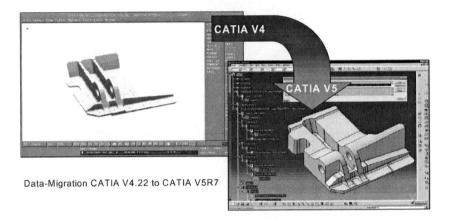

Data-Migration CATIA V4.22 to CATIA V5R7

Figure 3: Interpretation mistakes throughout migration of native CAD model information

That is, the approach of archiving native formatted product information provides no consistent solution.

3 Project group "AECMA-STAN LOTAR"

The objectives of the working group belong to the development of an AECMA Standard EN9300-x series supporting methods, processes and data models for LOng Term Archiving and Retrieval.

3.1 Objectives of project LOTAR

The intention of LOTAR is to enable Long Term Archiving of CAx and PDM data by providing methods, reference process and data models. The main priority is the fulfillment of certification and product liability requirements, which enables certified and auditable processes. Further priorities are the reuse of archived data and turning away from 2D drawings. The development of recommendations for industrial introduction and application pilots is also within the project scope. Experiences have been gained with pilot implementation of participating companies for practical turning over of LOTAR results. The entire project will be submitted as European and international standard.

3.2 Structure of AECMA-STAN Standard EN9300-x

Because of the highly complex topics and the huge amount of information the standard series is divided into several subparts. The following figure shows the overview of EN9300 standard series

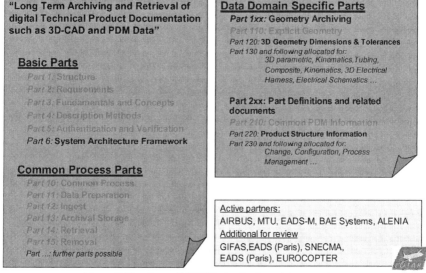

Figure 4: Structure of EN9300 standards and documents

The basic parts provide information about the common circumstances, descriptions of requirements, used and defined methods as well as information about verification and authentication. Furthermore descriptions of recommended and necessary system architecture may follow.

The Common Process Parts define the process steps necessary for the ensured provision and transfer of digital product information into and out of the archive.

The Data Domain Specific Parts offer information about, e.g., data quality and rules for CAD models and PDM information as well as validation properties, which ensure the consistency of the data throughout the procedures of LOTAR.

3.3 International Activities and Harmonization

It is a requirement of the entire aircraft and aerospace industry to harmonize the contents of EN9300 series of the AECMA-STAN LOTAR working group with the American regulation authorities under the umbrella of the International Aviation Quality Group (IAQG). The working title for the harmonization activity is LOTAR@AESPACE.

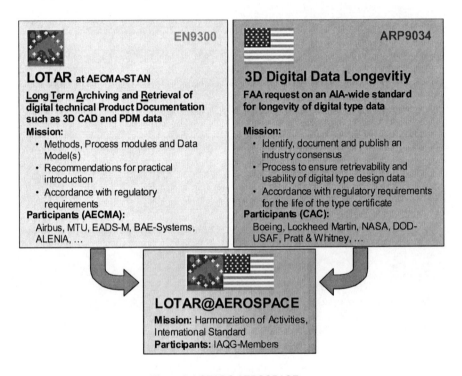

Figure 5: LOTAR@AEROSPACE

4 Project group „Langzeitarchivierung (LZA)"

Because of similar problems within the automotive industry several German OEM´s and 1st tier suppliers started a standardization group under the umbrella of the VDA research group "AK CAD/CAM".

The LZA project group aims for the development of recommendations for Long Term Archiving of 3D CAD product data and deals with the safeguarding of legal aspects of product liability and traceability as well as the reproduction of the data after years,

e. g., for spare part business. The VDA Guidelines will include descriptions of processes for Data Preparation, Archiving und Reuse:

Part 1: Overview, Requirements and general Recommendations

Part 2: Use Cases and LZA Processes

Part 3: LZA relevant Data and Models

Part 4: LZA Architectures

Part 5: Certification, Validation and Test

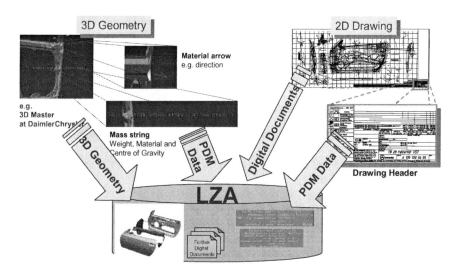

Figure 6: Long Term Archiving of 2D, 3D and PDM Data

5 Solution Approaches

5.1 Processes and Data of Project "LZA"

First of all product data are converted into standardized, "neutral" formats like ISO10303 (STEP), ASCII or TIFF, which are suitable enough for a long period in time. The period of archiving is normally more than 15 years with the duty of data upward compatibility and an independence of system and vendors over years.

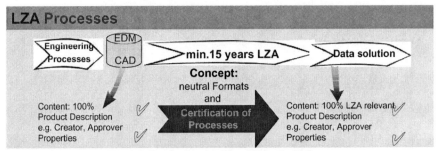

Figure 7: Fundamentals of Long Term Archiving

The second fundamental aspect is the definition and implementation of processes which can be certified to safeguard the unchangeability and authenticity of the product definition data.

A 3-level approach is recommended by the project group "LZA" according Figure 8 starting with the identification of the engineering data, which are relevant for long-term preservation and following with the description and mapping onto a neutral representation based on standards.

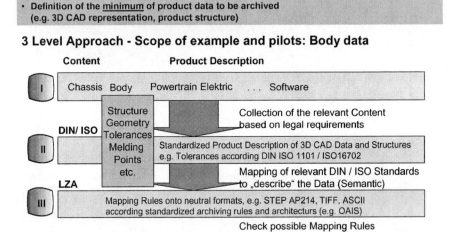

Figure 8: The 3-Level approach for archiving data definition

5.2 The basic fundamentals of LOTAR

LOTAR is based on international established standards to support processes, data formats and definitions. The main standards are:

- ISO103030 (STEP) for the definition of the neutral archiving data format
- ISO14721 (OAIS) for the definition of processes, requirements and system architecture

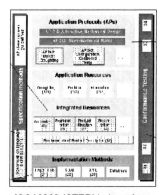

ISO10303 (STEP) to transform real (product) data descriptions into a neutral format for long term retention without loss of semantic

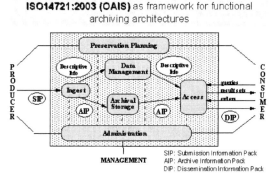

Digital Signatures according German digital signature law for unchangeability of data

Figure 9: Standards supporting LOTAR

5.3 Process Overview

Based on the functional architecture framework as recommended in ISO14721 the LOTAR project defined the phases and process steps shown in the overview in

Figure 10. The process description ensures the consistency of the data to be archived.

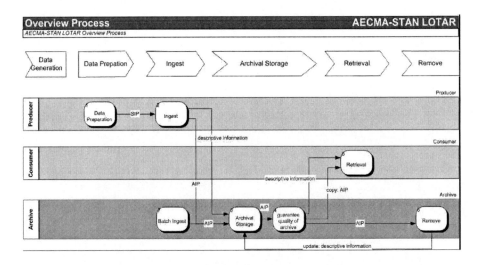

Figure 10: LOTAR Process Overview

The process is organized along the life cycle of the data to be archived. For this the data Producer has to prepare the data. After Data Preparation the resulting Submission Information Package (SIP) will be send during an Ingest Process to the Archive System. Within the Archival Storage Process the SIP will be converted into an Archival Information Package (AIP). This includes the conversion of native formatted CAD and PDM data as well as the signing of the data with a digital signature. In cases of reuse or viewing the data the Consumer has the possibility to retrieve the archived data. The archiving system may provide the optional Remove Process step in addition.

5.4 Data Preparation safeguarding data quality and content

The Data Preparation phase is essential for the success of the entire LOTAR process. A major problem belongs to the data quality during archiving. It is necessary, that the data quality shall be within defined parameters to ensure the readability within the next decades in various systems at the consumer site. LOTAR defines an explicit process which belongs to the data quality and data definitions (

Figure 11).

Long Term Archiving and Retrieval (LOTAR) of Digital Product Data

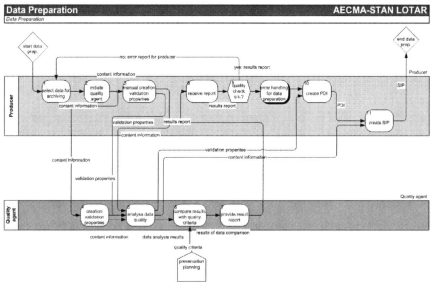

Figure 11: Data preparation process

It is important to produce data quality before archiving, because at the time of Ingest it is not known in which system the data will be viewed again or reused after years.

Therefore data quality improvement, e.g., of CAD models have to be performed before the archiving process can be initiated.

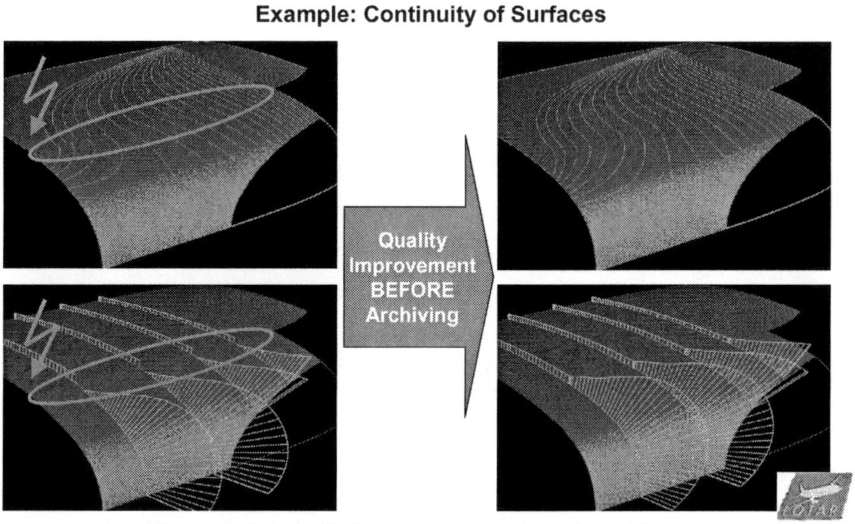

Example: Continuity of Surfaces

Figure 12: Data quality improvement necessary for archiving

The data quality checks as well as the monitoring of possible conversion procedures during the data flow into and out of the archive will be supported by Validation Properties. For example, the Validation Properties of a 3D CAD model will be generated 1st during the data generation and 2nd within the archive after the conversion into the correct archiving representation. A comparison of both values results in the acceptance or rejection of the data.

Because archiving problems are caused by „improper" 3D CAD models LOTAR recommends to define:

- The archive data model as unambiguously as possible
- Specific quality checks for the accuracy of model attributes BEFORE archiving the model
- Global checks of the source geometry and the archive format (e. g. volume, centre of gravity)
- A quality gate process ensuring those checks and the unchangeability of the data content throughout archive lifetime

It is recommended to follow the same approaches for PDM data, e.g., by Validation Properties for non-geometrical data.

5.5 Incremental Archiving & Retrieval

The archiving of digital product data shall be consistent with the workflow usually of the design procedures. This means, that an entire aircraft or car will not be designed within one single model but by thousands of models which are organized within a product (assembly) structure such as implemented in the 2D world by a huge set auf drawings. This results in the following statements for an archiving approach:

- Archiving and retrieval should be done only of a subset of product data at a certain time usually containing the relevant product data subset of interest
- Each subset of product data should be „complete" in itself for an unambiguous understanding
- The usage of product data subset provides the reduction of the amount of data to be archived depending on the scope of data for a concerned business process

It can be said that Long Term Archiving can be compared with Data Exchange, but without the possibility of a direct communication between Producer and Consumer.

Long Term Archiving and Retrieval (LOTAR) of Digital Product Data

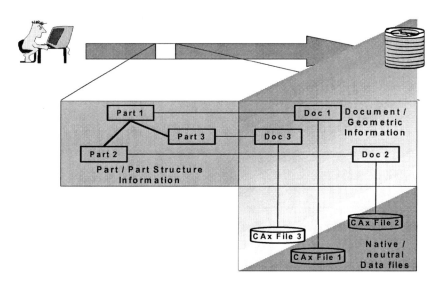

Figure 13: Submission of a complete product data subset

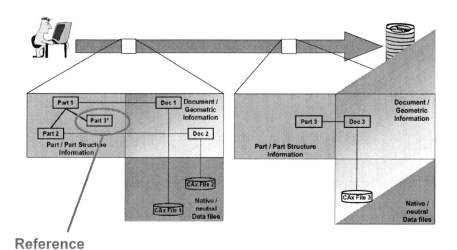

Reference
onto a component, which was already archived

Figure 14: Incremental archiving approach

6 Benefits of Reference Models for Long Term Archiving and Retrieval

The new trends in product modeling practice based on 3D CAD models and PDM structural data in combination with simplified drawings or without any 2D drawings lead to a lack of reliable archiving processes and tools. But the retention of product data and digital documents according standardized reference models for processes and data will enable

- Systematic archiving of the company know-how
- Availability of documents via a logically central instance
- Minimization of individual errors, e. g., through document versions, which are not of actual status
- High grade of security by a secure repository and access control
- Internal legitimating of digital processes currently in use
- Economic archiving of voluminous data
- High grade of security against attack, sabotage etc

That is, holistic product development has to include also the preservation of the engineering knowledge and methods as well as the retention and accessibility onto unambiguous product definition data and descriptions (schema) for more than 15 years in an appropriate quality. Furthermore it has to fulfil the requirements for archival, retrieval and use of digital data, which are necessary to meet business, regulatory and legal conditions.

Verifiable and auditable LOTAR Processes, the identification of the proper incremental data content (in the correct quality, representation and format) in combination with the proof of unchangeability and invariance over years, e.g., by using digital signatures are key aspects for long term archival solutions of 3D CAD and PDM data.

7 References

[1] ISO 10303-214; Industrial automation systems and integration - Product data representation and exchange: Part 214 Application Protocol: Core data for automotive mechanical design processes
[2] ISO 14721.2003; OAIS: Open Archiving Information Model – Reference Model
[3] Reference Model for an Open Archival Information System (OAIS): Foreword, CCSDS 650.0-B-1, BLUE BOOK, January 2002

8 Authors

A. Trautheim, PROSTEP AG, Dolivostraße 11, 64293 Darmstadt,

Andreas.Trautheim@prostep.com

L. Millarg, PROSTEP AG, Dolivostraße 11, 64293 Darmstadt,

Ludger.Millarg@prostep.com

Part V: *Fundamentals*

Surface Design through Modification of Shadow Lines

Roger K. E. Andersson

Mathematical Sciences, Chalmers University of Technology and Göteborg University, Göteborg, Schweden

Abstract: A shadow line in a surface consists of the points in the surface at which the normal is perpendicular to the direction vector of a distant light source. Families of such lines are extensively used in the analysis of surface shape.

Currently, they are used in a trial-and-error process: the surface designer displays them, tries to conceive how to modify the surface to improve them, changes the surface accordingly, and displays the lines of the modified surface, in a typical design cycle. The cycles are repeated until the lines are considered good enough.

In the paper, we will sketch how this extremely time-consuming process may be replaced with a direct process in which the surface designer indicates a desired pattern of shadow lines while a surface having this pattern of shadow lines is generated automatically through mathematical software.

1 Introduction

The subject of this paper originated in the project FIORES (**F**ormalization and **I**ntegration of an **O**ptimized **R**everse **E**ngineering **S**tyling Workflow) funded by the European Commission, and coordinated by Prof. C. W. Dankwort. A main theme of this project was Engineering in Reverse, EiR for short, which in particular amounts *Modelling with properties instead of data*. One important way to realise it is to *enable users to set up properties an object (curve or surface) should possess, then find the* *desired object through mathematical software.* In this paper, *we will outline what such a process may amount in the case of shadow lines.*

We start in Chapter 2 by looking at what shadow lines are, and in particular, how one may set up families of them to cover a portion of the surface of interest and to enable surface design through their modifications. In Chapter 3, we show that shadow lines may be regarded as level curves of a real valued function defined in the parameter domain of the surface, and lifted to the surface by its equation. Thought simple, this is indeed a key observation behind the process.

In Chapter 4, we briefly discuss some basic problems akin to editing shadow lines and in Chapter 5 we interpret the result of modifications of families of shadow lines in our setting. In Chapter 6, containing the main results of the paper, we show how to find a surface having the modified shadow lines and discuss what further requirements are needed to get a unique solution. It also shows that even if being highly desirable, no further requirements can be put, because they rule out the possibility to find a solution. In the concluding Chapter 7, we present the surface obtained, and consider some of it properties. In particular, we look at how well its shadow lines agree with the ones imposed.

Much of the machinery needed to carry out the suggested plan will inevitably rely on mathematics and numerical analysis. We have tried to keep these matters to a minimum. In a few cases, we have not been able to explain what is going on without using mathematical expressions, but then endeavoured to illustrate the meaning of these expressions, in general by using graphics from Mathematica. Indeed, for the purpose of exhibiting what the process amounts, it has been implemented in Mathematica in a simple case, and all illustrations in the figures are the result of real computations in Mathematica. These efforts to reveal what is going on are the reason for the numerous figures that furnish an important part of the presentation.

2 What shadow lines are, and how to set them up

In the following the surfaces, often denoted by S, are assumed to be parametric surfaces. Often, the equation of the surface is denoted by $\mathbf{r}$, which is a function of two variables u, v. A generic point P in the surface is given by $P = \mathbf{r}(u, v)$, where (u, v) is a point in some domain D in the plane, the parameter domain of the surface. Such a surface is illustrated in Figure 1, in which it is a simple bicubic spline, and the parameter domain is rectangular.

Figure 1: The original surface.

A shadow line in a surface consists of the points in the surface at which the normal is perpendicular to the direction vector of a distant light source. The next figure gives an example. Here, we have a point P_0 in the surface, its upward unit normal $\mathbf{n}_0$, a direction vector $\mathbf{q}_0$ perpendicular to the normal, which is the direction vector of a far distant light source, and the curve in the figure is the corresponding shadow line. The point P_0 lies on this shadow line, because the normal of the surface at the point is perpendicular to the light source direction, and the same applies for all points of the shadow line. Now we have one shadow line, but it alone does not tell so much about the shape of the surface. Instead, we need a family of shadow lines, spread out over the part of the surface of interest, which we for simplicity assume is the entire surface. To set up these, we will use the unit vector $\mathbf{p}_0$ in the vector, pointing upwards. Let $\mathbf{q}_s = \mathbf{q}_0 + s\mathbf{p}_0$ and take $\mathbf{q}_s$ as the direction of light source. For each s, we will in principle get a shadow line, i.e. the one consisting of all the points in the surfaces, at which the normal is perpendicular to $\mathbf{q}_s$.

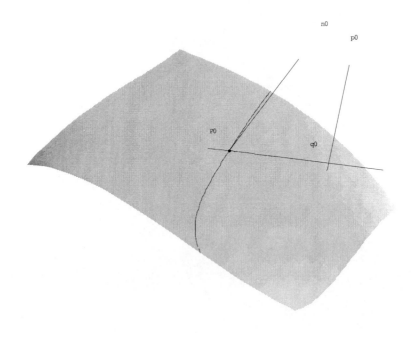

Figure 2: The setup for shadow lines.

As will easily turn out of the figure, not all numbers s will give rise to shadow lines. If we take s positive and large, or s negative and large, there will be no point in the surface at which the normal is perpendicular to $\mathbf{q}_s$. It is not hard to see that the s-values giving rise to shadow lines in the surface are the ones that satisfy $s_{\min} \leq s \leq s_{\max}$ with $s_{\min}$ and $s_{\max}$ the minimum and maximum, respectively of the expression

$$r(u,v) = -\frac{\mathbf{q}_0 \cdot \mathbf{n}(u,v)}{\mathbf{p}_0 \cdot \mathbf{n}(u,v)} = -\frac{\mathbf{q}_0 \cdot \mathbf{r}_u(u,v) \times \mathbf{r}_v(u,v)}{\mathbf{p}_0 \cdot \mathbf{r}_u(u,v) \times \mathbf{r}_v(u,v)},$$

where the second equality follows from $\mathbf{n}(u,v) = \frac{\mathbf{r}_u(u,v) \times \mathbf{r}_v(u,v)}{|\mathbf{r}_u(u,v) \times \mathbf{r}_v(u,v)|}$. In these formulas, the dot denotes scalar multiplication and the cross denotes the cross product of vectors and $|\mathbf{w}|$ denotes the length of the vector $\mathbf{w}$.

3 A key observation behind the process

As a background, we start by considering how to find the shadow line corresponding the vector $\mathbf{q}_s = \mathbf{q}_0 + s\mathbf{p}_0$ for some given s-value. **This is a key observation behind our process.**

From the definition of shadow lines, the parameter value $(u,v) \in D$ has the property that the point $P = \mathbf{r}(u,v)$ lies on the shadow line corresponding to the value $s \Leftrightarrow$

$$\mathbf{n}(u,v) \cdot \mathbf{q}_s = 0 \Leftrightarrow \mathbf{n}(u,v) \cdot (\mathbf{q}_0 + s\mathbf{p}_0) = 0 \Leftrightarrow \mathbf{n}(u,v) \cdot \mathbf{q}_0 + s\mathbf{n}(u,v) \cdot \mathbf{p}_0 = 0 \Leftrightarrow -\frac{\mathbf{n}(u,v) \cdot \mathbf{q}_0}{\mathbf{n}(u,v) \cdot \mathbf{p}_0} = s.$$

Thus, recalling that we have introduced the function r defined in the parameter domain D through

$$r(u,v) = -\frac{\mathbf{q}_0 \cdot \mathbf{n}(u,v)}{\mathbf{p}_0 \cdot \mathbf{n}(u,v)} = -\frac{\mathbf{q}_0 \cdot \mathbf{r}_u(u,v) \times \mathbf{r}_v(u,v)}{\mathbf{p}_0 \cdot \mathbf{r}_u(u,v) \times \mathbf{r}_v(u,v)},$$

in Chapter 2, it follows that the shadow lines of the surface S are nothing but the images under $\mathbf{r}$ (the function representing the surface) of the level curves of the function r. We have not assumed any special property of the surface for this relation to hold, so it holds for any surface. We also note that from the second formula for r it follows that if $\mathbf{r}$ is a spline surface then r is a piecewise rational function.

4 Editing shadow lines

To make reliable modifications to the shadow lines, one must select a view in which they are all displayed well as the working view. For reasons that will turn out later, one should keep this view as the working view, i.e. the only view in which one imposes changes to the shadow lines. For our example surface, we have chosen the top view in Figure 3, where we see the 25 shadow lines that result by taking for the s introduced in Chapter 2, 25 values evenly distributed between s_{min} and s_{max}.

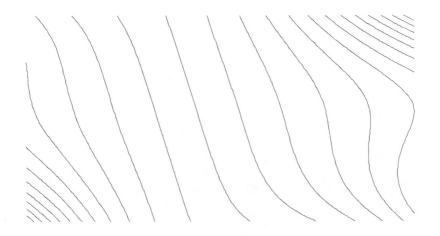

Figure 3: Shadow lines seen from above.

Picking a shadow line and reshaping it according to a designer's intension seems to be a natural way to impose modifications to shadow lines There are several complications, however, making it much harder to come up with good working tools.

First, one must keep in mind that it is impossible to change just a single shadow line. As soon as one is designing a smooth surface, whatever change one imposes on a shadow line will imply changes of shadow lines in its neighbourhood, when the surface is modified according to the demands. Indeed, the problem concerns how to edit a **family** of curves in a convenient way, not a single one at the time.

One method that is currently being studied is to enable change of any of the displayed curves and letting the modification decay away from the curve in a manner that the user may govern and superimpose the changes of the individual curves into a total change. Though promising, it is too early to draw far reaching conclusions for the moment, and we feel that one should consider the area as one where more research is needed.

Concerning modification of shadow lines, one should also keep smoothes to adjacent surfaces, if any, in mind. To maintain G^k – continuity along some part of the boundary, any modified shadow line meeting the part must coincide in G^{k-1}-sense with the corresponding unmodified shadow line at the intersection with the boundary.

Here, we say that two curves coincide in G^m-sense at a point provided that

- they coincide at the point for $m = 0$

- the curves and their unit normals coincide at the point for $m = 1$

- the curves, their unit normals and their radii of curvatures coincide for $m = 2$.

Figure 4 shows a modification of this kind, where we want to keep a G^2 – continuity along the lower- and the left edges of the rectangle. It may perhaps be said that this is a large modification, with the purpose to show that the underlying machinery is by no means confined to small modification, contrary e.g. to the case of modification based on isophotes or reflection curves.

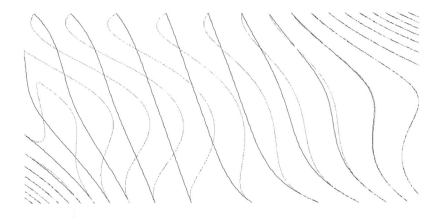

Figure 4: Original and desired shadow lines.

The next figure shows the desired shadow lines alone:

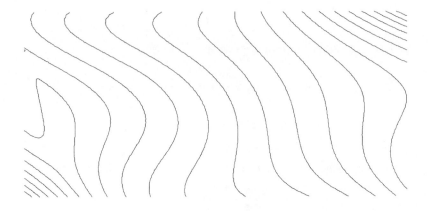

Figure 5: The desired shadow lines.

5 An interpretation of the changing of shadow lines

Recalling the way we have related shadow lines in the surface with level curves of a function defined in the parameter domain of the surface in Chapter 3, the effect of modifying the entire family of reflection lines may be considered as a modification Δr of this function. Thus the desired shadow lines, of which some are shown in Figure 5, correspond to level curves of the function $r + \Delta r$, defined in the parameter plane.

Figure 6 shows the function r corresponding to the original shadow lines shown in Figure 3, whereas Figure 7 shows the modification Δr corresponding to the modification of these lines, shown in Figure 4.

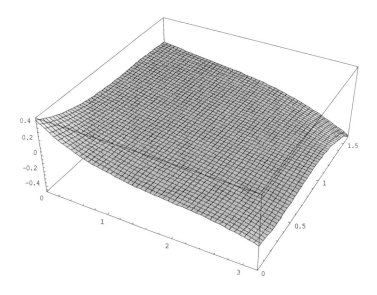

Figure 6: The function r over the parameter plane,
inducing the original shadow lines.

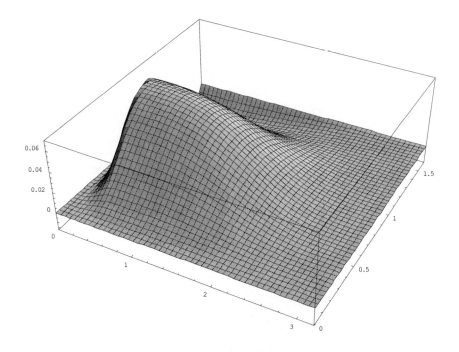

Figure 7: Change Δr of r caused by the modification of shadow lines.

6 Finding the surface having the modified shadow lines

Now consider a new surface S', with the same parameter domain D as that of S, and with equation $\mathbf{s}(u,v) = \mathbf{r}(u,v) + h \cdot \mathbf{d}$, where h is a real valued function in D and $\mathbf{d}$ is a unit vector. Thus S' may be called a variable offset surface with offsets along the direction $\mathbf{d}$, and where the size of the offset from the surface S at the point $\mathbf{r}(u,v)$ is given by $h(u,v)$. We are going to determine the function h in such a way that the surface S' will have the desired shadow lines.

Since all offsets from the surface S are along the direction $\mathbf{d}$, as long as we look at the surfaces in the $\mathbf{d}$–direction, we cannot see any differences between S and S'. In particular, it does not matter whether we carry a curve in the parameter domain through the function $\mathbf{r}$ or the function $\mathbf{s}$. Thus, we may transport changes imposed

Surface Design through Modification of Shadow Lines

on the shadow lines **in this view** to the parameter domain with $\mathbf{r}$. Likewise, we may lift level curves to the function

$$s(u,v) = -\frac{\mathbf{s}_u(u,v) \times \mathbf{s}_v(u,v) \cdot \mathbf{q}_0}{\mathbf{s}_u(u,v) \times \mathbf{s}_v(u,v) \cdot \mathbf{p}_0}$$

with the function $\mathbf{r}$ and state that they are the shadow lines to S' **when seen in this view**.

The importance of this remark is that strictly speaking, we should use $\mathbf{s}$ for both of these purposes. But $\mathbf{s}$ is unknown until h is determined while $\mathbf{r}$ is known all the time! We have already suggested how the user may set up the function $s = r + \Delta r$ through modification of one or several shadow lines. For the moment, we assume that this has already been done, thus s is a known function. Since the changes to the shadow lines are confined to the interior of the rectangle D, the function s coincides with the function r on the boundary of D.

We will now see how to choose the function h in such a way that

$$\frac{\mathbf{s}_u(u,v) \times \mathbf{s}_v(u,v) \cdot \mathbf{q}_0}{\mathbf{s}_u(u,v) \times \mathbf{s}_v(u,v) \cdot \mathbf{p}_0} = -s(u,v).$$

From $\mathbf{s}_u = \mathbf{r}_u + h_u \cdot \mathbf{d}$ and $\mathbf{s}_v = \mathbf{r}_v + h_v \cdot \mathbf{d}$ follows

$$\mathbf{s}_u \times \mathbf{s}_v = \mathbf{r}_u \times \mathbf{r}_v + h_v \cdot \mathbf{r}_u \times \mathbf{d} + h_u \cdot \mathbf{d} \times \mathbf{r}_v.$$

Inserting this in the previous expression and simplifying, we get the equation

$$\mathbf{d} \times \mathbf{r}_v(u,v) \cdot (\mathbf{q}_0 + s(u,v) \cdot \mathbf{p}_0) h_u(u,v) + \mathbf{r}_u(u,v) \times \mathbf{d} \cdot (\mathbf{q}_0 + s(u,v) \cdot \mathbf{p}_0) h_v(u,v) =$$
$$\mathbf{r}_u(u,v) \times \mathbf{r}_v(u,v) \cdot (\mathbf{q}_0 + s(u,v) \cdot \mathbf{p}_0).$$

Thus, with

$$A(u,v) = \mathbf{d} \times \mathbf{r}_v(u,v) \cdot (\mathbf{q}_0 + s(u,v) \cdot \mathbf{p}_0),$$
$$B(u,v) = \mathbf{r}_u(u,v) \times \mathbf{d} \cdot (\mathbf{q}_0 + s(u,v) \cdot \mathbf{p}_0),$$
$$C(u,v) = \mathbf{r}_u(u,v) \times \mathbf{r}_v(u,v) \cdot (\mathbf{q}_0 + s(u,v) \cdot \mathbf{p}_0),$$

we get the equation

$$A(u,v)h_u + B(u,v)h_v = C(u,v).$$

This is a linear first order hyperbolic partial differential equation for h that we must solve to produce the surface having the shadow lines the user is asking for. It is an equation of a kind well known in particular in fluid dynamics. The correct boundary conditions for such an equation, i.e. the conditions under which one can find a unique solution have been known during at least the preceding 40 years, possibly even longer and run as follows.

Let β be the vector whose components are the coefficients for h_u and h_v respectively, i.e. $\beta(u,v) = (A(u,v), B(u,v))$ and let $v(u,v) = (U(u,v), V(u,v))$ be the exterior normal to the boundary ∂D of D. Then the **inflow boundary** is the set

$$B_- = \{(u,v) \in \partial D \mid v(u,v) \text{ exists and } \beta(u,v) \cdot v(u,v) < 0\}$$

and the **outflow boundary** the set

$$B_+ = \{(u,v) \in \partial D \mid v(u,v) \text{ exists and } \beta(u,v) \cdot v(u,v) > 0\}$$

Since $s(u,v) = r(u,v)$ on ∂D we have

$$\beta(u,v) = (\mathbf{d} \times \mathbf{r}_v(u,v) \cdot (\mathbf{q}_0 + r(u,v) \cdot \mathbf{p}_0), \mathbf{r}_u(u,v) \times \mathbf{d} \cdot (\mathbf{q}_0 + r(u,v) \cdot \mathbf{p}_0))$$

on ∂D, hence everything needed to determine B_- and B_+ is known as soon as $\mathbf{d}, \mathbf{q}_0, \mathbf{p}_0$ are known. Thus when choosing the working view direction, light source direction and the unit vector used for defining the set of light sources, one determines also which part of the boundary to fix and what to leave free. For this reason, it is appropriate to inform the user on the behaviour of $\beta \cdot v$ when setting them up, e.g. by displaying this function over the boundary of the parameter domain, like in Figure 9. The correct setting of boundary values is to put $h = 0$ on B_- and let h be free on B_+ or vice versa. Figure 8 shows a correct setting in the present case.

Surface Design through Modification of Shadow Lines

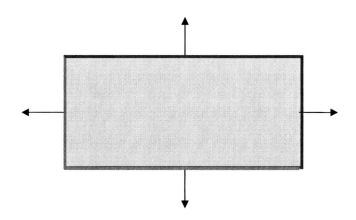

Figure 8: Inflow- and outflow boundaries determined by the sign of $\beta \cdot v$ (red < 0, blue > 0).

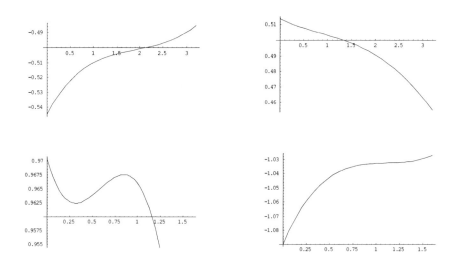

Figure 9: $\beta \cdot v$ over the four edges of the rectangle in the parameter plane.

7 The resulting surface and its properties

Solving the boundary value problem presented in the preceding chapter, we obtain the surface that we are looking for, i.e. the surface whose pattern of shadow lines we have prescribed. In the actual case, we have used NDSolve in Mathematica, using its MethodOfLines as method for solving the boundary value problem, and with $h = 0$ along the lower- and left edges of the rectangle, marked by red in Figure 8.

The resulting surface is shown in Figure 10. It is not easy to see any difference between it and the surface we commenced with in this scale. Figure 11 displays the vertical distances between the resulting surface and the one we started with, exhibiting a difference with a maximum of about 0.08. It may seem quite small, taking the large effects on the shadow lines into account. It must be related to the size of the surface, and such a comparison tells that the maximum deviation corresponds to about 20 mm for a surface of length 400 mm times 800 mm, and related to common orders of magnitude in automotive design, such a deviation is not small.

Figure 12 shows the shadow lines of the new surface. In Figure 13, the desired shadow lines, set up in Chapter 4 are drawn in red, together with the shadow lines of the computed surface. It is hard to see any difference. To get such a good agreement, the boundary value problem must be solved rather accurately.

From Figure 7, it turns out that the largest modification of shadow lines occur close to the middle of the left hand side of the surface, whereas Figure 11 tells that the largest modification of the surface does occur in the vicinity of the upper and right edge. At a fist glance, it may appear remarkable, but it is quite in order and is a reason why one cannot fix the entire boundary and still be able to modify the shadow lines the way one may want. The modification of shadow lines indicated in Figure 7, interpreted as a change of the function r is positive thorough, and does not allow for moving the surface in opposite direction.

Surface Design through Modification of Shadow Lines

Figure 10: The new surface.

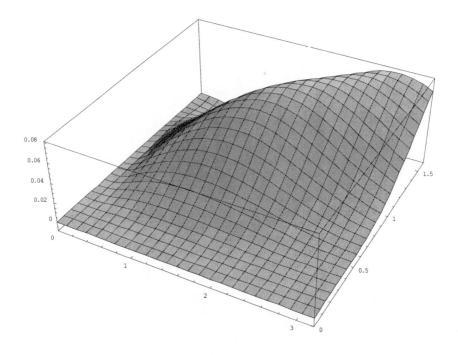

Figure 11: Difference new – old surface.

Figure 12: Shadow lines of the new surface.

Surface Design through Modification of Shadow Lines

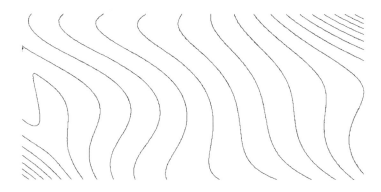

Figure 13: Desired shadow lines (red), the ones of the computed surface (black).

We conclude by visualising the deviation between the new and the old surfaces along the upper boundary and right boundary in Figure 14 and Figure 15 respectively.

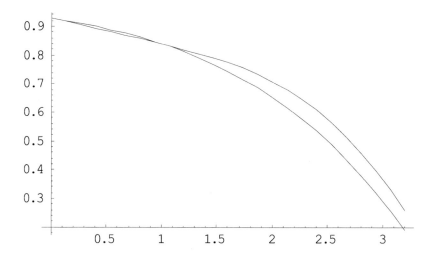

Figure 14: Deviation along the upper boundary.

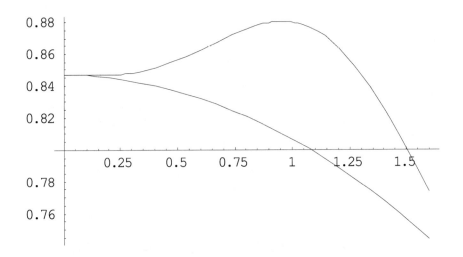

Figure 15: Deviation along the right boundary.

8 Author

Roger Andersson, Chalmers University of Technology and Göteborg University,

Eklandagatan 86, SE 41296,

rogand@math.chalmers.se

Representation of Intersection Curves

Roger K.E. Andersson [1]; Bo I. Johansson [2];

1) Mathematical Sciences, Chalmers University of Technology and Göteborg University, Göteborg;
2) Mathematical Sciences, Chalmers University of Technology and Göteborg University, Göteborg.

Abstract: In most current CAx systems one may easily find a B-spline curve that approximates the intersection of two B-spline surfaces with a desired accuracy. As a part of an ongoing project on CAD quality, we have studied some important properties of such curves obtained within several common CAx systems. The resulting curves exhibit considerable variations that are presented and discussed. We also present a couple of procedures that for common kinds of B-spline surfaces generates approximations to their intersection curves with a desired accuracy, with an almost minimal complexity and that adhere to the smoothness of the surfaces.

1 Introduction

The purpose of this paper is to consider some CAD quality problems that show to be of highly practical importance and to propose remedies that can immediately be implemented with little efforts and causing only minor changes to current CAx systems.

We start in Chapter 2 by reviewing some of the efforts that have been paid by groups of users to enhance the situations that commenced in a public level about a decade ago, that we have been involved with during the last three years and that have recently led to global guidelines. Prior to it, we are aware on efforts on CAD quality at individual companies within the automotive industry, often leading to in-house software to check if a CAD model was affected by some common errors or not.

Almost all work on CAx quality initiated by a group of users targets the role of the user, giving recommendation on how to enhance the quality. While agreeing on the importance of this work, it was soon clear to us that this work alone would not be enough to efficiently avoid defects. Indeed, analysing actual errors, we found that a good deal of them emanates not from an improper handling by a user but **by weakness in the software**.

In this paper, we concentrate on one of these, namely that of representing intersection curves between surfaces. In the second part of Chapter 2, we present a brief account of why this is a problem and indicate a way to milder its effects. In Chapter 3, we describe a few simple test surfaces that we have been using to asses the representation of intersection curves in a large number of current CAx systems, and in Chapter 4, we indicate the results of the tests of a small selection of systems, presented in an anonymous manner.

These results are contrasted with the results obtained with the methods we propose and that we describe in Chapter 5. Our methods were implemented in Open Cascade 5.0, which was also used as the software platform for parts of the evaluations.

We are happy to acknowledge the usefulness of the very extensive CAD software Open Cascade in public domain, with open source code, an invaluable source for testing and rapid implementations of new ideas and methods within computer-aided design.

2 Motivation of the study

During the last decade there has been an increasing user-driven interest in reducing CAD quality problems. An early contribution of guidelines with the purpose to increase the quality of CAD models was made available in 1993 by the organisation Verband der Automobilindustrie (VDA) in Germany through the introduction of Version 1 of VDA-Empfehlung 4955, with the name "Qualität und Umfang von CAD/CAM-Daten". It was an encounter of defects of CAD descriptions that in practice had proved to often prevent their use and raising the need of rework. An elaborated and augmented in Version 2 replaced this recommendation in 1999.

The mean effects of insufficient CAD quality were early recognised in the automotive industry world-wide and individual programs to milder its influences were established. In 1999, the problem was realised to be common to the international automotive industry and that it was best treated by joint efforts, leading to the formation of the organisation **SASIG-PDQ,** which is the acronym for S̲trategic A̲utomotive product data S̲tandards I̲ndustry G̲roup. Currently it consists of the following organisations:

- Automotive Industry Action Group, AIAG, (U.S.)
- Federal Chamber of Automotive Industries, FCAI, (Australia)
- Groupement pour l'Amélioration des Liaisons dans l'Industrie Automobile, GALIA, (France)
- Japan Automotive Manufacturers Association, JAMA-JAPIA, (Japan)
- ODETTE Sweden, (Sweden)
- Verband der Automobilindustrie, VDA, (Germany).

Based on VDA-Empfehlung 4955, Version 2, and similar documents from the other organisations, one started the work to generate global guidelines for data quality. In March 4, 2004, SASIG Product Data Quality Guidelines for the Global Automotive Industry, Version 2, was consensus approved for release. For further information, we refer to the relevant national organisation in the above list.

In a local level, the research project FUNCAD, financed by Swedish Agency for Innovation Systems, was considering the CAD quality problem from a somewhat different point of view. While actively participating in SASIG-PDQ since 2000, representing Odette Sweden in the development of the global guidelines, thereby gaining a good overview over the totality of problems encountered in the practical work, the main research objective was to identify an important source of the problems and attack it.

It was soon realised that **surface intersections is an outstanding single creator of CAD quality problems.** The extent of the problem derives from the current de facto standard for describing CAD surfaces as a network of trimmed NURBS. Having been an exceptional solution up to about the end of the 80-s, it has become the rule today. In addition, the increased use of solid modelling with parametric free-form surfaces forces even more surface-surface intersection to be done, stressing the problem further. It is interesting to notice that this conclusion seems to be widespread among mathematicians working in the field. A consensus opinion among the participants of the Workshop on Mathematical Foundations of CAD, held at Mathematical Science Research Institute, (MSRI), Berkeley, California, in June 4-5, 1999 is:

"The single greatest cause of poor reliability of CAD systems is lack of consistent surface intersection algorithms"

A main basic reason behind has been known among researchers within Computer Aided Geometric Design (CAGD) for many years: **intersection curves cannot in general be represented exactly in CAD systems.** As was shown by Katz and Sederberg [1], already in 1988, the curve of intersection between two bi-cubic patches in general position is a space curve of degree 324 of genus 433 and, because its genus is non-zero, cannot be represented as a rational Bezier curve of any degree.

To enable handling in a CAD system, the true intersection curves must be approximated by the kind of curves available in the system, which today means NURBS, or in practice, some subclass of NURBS. In most current systems, intersection curves are represented as non-uniform B-spline curves of moderate degree.

Before we discuss the problem of finding suitable approximations of this kind, we briefly indicate some inevitable consequences of the approximation per se, as these are reasons for difficulties encountered in the use of CAD models. Figure 1 shows an intersection between two B-spline surfaces.

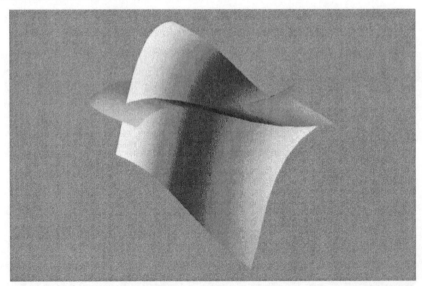

Figure 1: Two intersecting surfaces.

As a consequence of the fact that the curve used to represent the intersection is only an approximation, this curve **does not lie in any of the surfaces.** This fact is illustrated in a common CAD system in Figure 2.

Representation of Intersection Curves

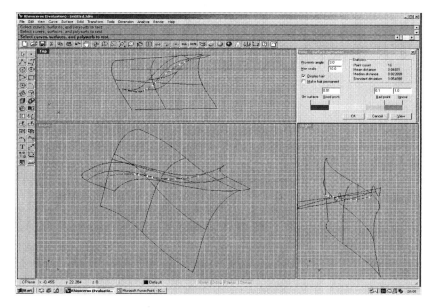

Figure 2: The « intersection » curve between the surface in Figure 1 and its distance to one of the surfaces, measured in a few points along the curve. The relative distance is suggested by the length of the coloured bar along the normal to the surface. The colour scale associates an interval to the colour.

A very common reason for finding the intersection between surfaces is its use in a trimming operation, i.e. an operation in which parts of the current surfaces are cut away, giving rise to a trimmed surface. Figure 3 shows the lower part of the lower surface in Figure 1 that remains after the part above the intersecting surface is removed.

In this operation, however, the space curve representing the intersection between the two surfaces is not used itself. Here, as indicated in Figure 4, the corresponding curve in the parameter plane of the surface is needed. This curve is determined in one way or another in such a way that when composed with the surface-map it gives an approximation of the true intersection curve. Thus we have indeed **three** curves representing the intersection curve, one curve in **each** of the two intersecting surfaces, lying in the surface being the composition of a curve in the parameter plane of the surface, and the space curve we described above.

261

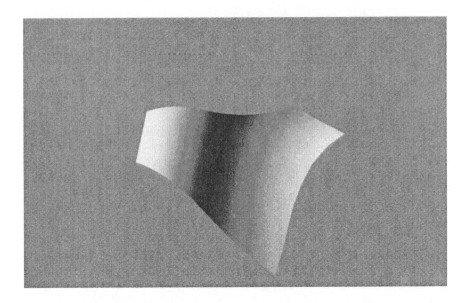

Figure 3: The trimmed surface.

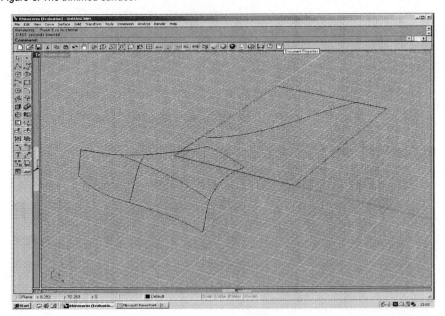

Figure 4: One of the « intersection » curves and the corresponding curve in the parameter plane of the surface.

Representation of Intersection Curves

These three representations of the intersection curve are all different! It can be made visible, at least in the cases when a coarse tolerance is being used, simply by zooming enough. It is illustrated in Figure 5, where a tolerance of 0.02 is being used. The yellow curve is the approximation coming from the parameter plane of one of the surfaces and the black curve is the space curve approximating the intersection curve.

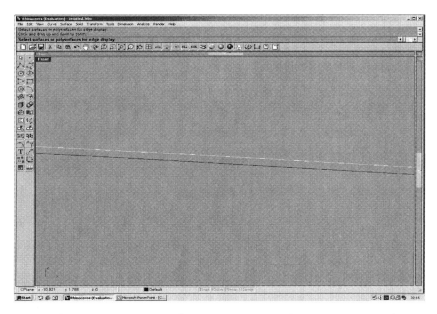

Figure 5: If one zooms enough, the different representations of the intersection curve fall apart.

The problem emanating from the approximations of intersection curves is hard, sometimes impossible for the user to affect:

- In some systems, the maximum allowed distance between the "intersection" curves is pre set. On conversion to a system using a more narrow tolerance, a gap will arise.

- If the distance may be set, one must consider a smaller distance against an increase in complexity of curve and surface descriptions. If leading to an excessive number of arcs, it may come in conflict with the SASIG-PDQ guidelines mentioned in Chapter 2.

- **Remedy:** *represent the intersection curve with as few arcs as compatible with the tolerance!*

3 The three test surfaces to intersect

Here, we will describe the kind of surfaces used in the tests. The C^2 – case, i.e. the case in which both the surfaces are C^2 is shown in Figure 6. The smaller surface, in the upper part of the figure, is a B-spline surface of u-degree 3 and v-degree 3 with u-knots 0 and 1, each of multiplicity 4, and v-knots 0, 1, 2, 3, 4, 5 where the knots 0 and 5 are of multiplicity 4 and the other (interior) ones are simple knots. Thus the surface is at least C^2. **This surface will be used as the second component of the two intersecting surfaces in all test.**

We now consider the first component of the intersecting surfaces, whose C^2 version is the bigger surface in Figure 6. We first describe its C^1 version. It is a B-spline surface of u-degree 1 and v-degree 5 with u-knots 0 and 1, each of multiplicity 2. The v-knots, with 4 correct decimals, are 0, 1, 1.5544, 1.9712, 2.1651, 2.4683, 3.2106, 3.4681, 3.6813, 4.8010 with 0 and 4.8010 of multiplicity 6 and the other of multiplicity 4. Thus the surface is at least C^1. It is **not** C^2, having a relative jump in radii of curvatures of more than 1.88 at the v-knot 3.6813 of the boundary curves u=0 and u=1. Here the relative jump of curvature is defined as $\dfrac{2|r_1 - r_2|}{r_1 + r_2}$ with r_1 and r_2 the radius of curvature immediately before and after the knot respectively. More precisely r_1 and r_2 is the left and right limit, respectively, of the radius of curvature when the v-parameter tends to the v-knot value.

To give an idea of the size of the surfaces, we indicate the coordinates of the four vertices of the bigger surface. Upper left: (0, 5, 5), upper right: (10, -45, 0), lower left: (-40, 0, 50), lower right: (-15, 45, 50), all in millimetres.

It is now easy to describe the C^2 version seen in Figure 6. It is simply the C^1- surface just described, but with the multiplicity of all interior v-knots lowered from 4 to 3, forcing the resulting surface to become C^2. This operation changes the surface with less than 1 mm, so one cannot see the difference of the C^1 and the C^2 versions in a figure like Figure 6.

Representation of Intersection Curves

The C^0 version is created from the C^1 version in the following way. First we increase the multiplicity of the knot 1.5544 from 4 to 5 and then move the control points of the surface to create an edge, with the result shown in Figure 7. From Figure 8, it clearly turns out that it results in an intersection curve that is far from tangent continuous.

Figure 6: The C^2 - surfaces used in the test.

Figure 7: The C^0 - surface and the C^2 - surface used in the test.

Representation of Intersection Curves

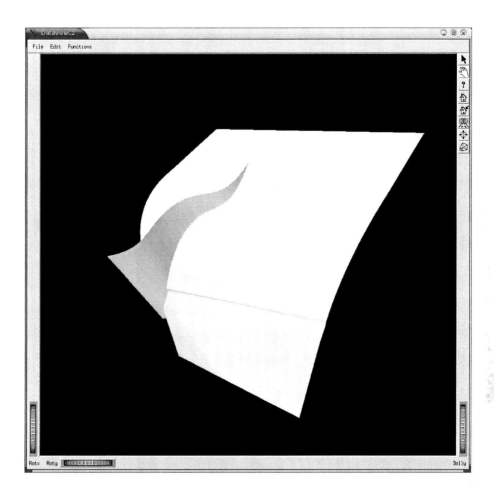

Figure 8: The C^0 - surface and the C^2 - surface seen from another viewpoint, indicating a considerable lack of G^1 continuity of the intersection curve.

4 Representation of the intersection curves in a few CAx systems

We will now present a sample of the results that we have obtained by creating intersection curves, using different tolerances, in a large number of current CAx systems. Here, the x in CAx seems quite appropriate, since the study comprises systems from ones often considered to particularly target the styling/design phase to ones considered to in particular target the engineering/design phase of product development. Also considering the cost of systems, the study covers a broad range from inexpensive to the top range in regard of cost.

In some of the cases, we contrast these results with the result we have obtained with our method that we will present in the next chapter. In three tables, we summarise results of:

- Intersections with the tolerances 0.02, 0.001 of the C^1 - surfaces and the C^2 - surface in four systems together with our C^2 - curves.
- Intersections with the tolerances 0.02, 0.001, 0.0001 and 0.00001of the two C^2 - surfaces in two systems together with our C^2 - curves.
- Intersections with the tolerances 0.02, 0.001, 0.0001 and 0.00001 of the C^0 - surface and the C^2 - surface in two systems.

The purpose with this study is to overview the current situation and to propose possible improvements, by no means to rank the investigated CAx systems. Therefore, the systems from which results are obtained are referred to as System X, where X is a capital from the beginning of the alphabet.

For readers interested to see how a particular system that they have access to behaves, we are ready to provide the surfaces used in the test as three IGES files on request by email to one of the authors.

In the tables, #arcs refers to the number of arcs in the curve, equivalently one less than the number of distinct knots, #ctrlp refers to the number of control points in the curve, whereas deg stands for the degree of the curve, equivalently one less than the order of the curve.

Moreover, cont means the degree of continuity of the curve, which is the least order of continuity of any knot, if it varies along the curve. Here, 0 means C^0, 1 means C^1

Representation of Intersection Curves

and 2 means C^2. Finally, maxdist1 is the maximum distance to the first mentioned surface and maxdist2 is the maximum distance to the second mentioned surface in the couple to intersect.

We remind that in all cases, the second surface in the couple is always the small surface described in the preceding chapter. The first surface is the version of the bigger surface described in the previous chapter, denoted as the C^0 -, C^1 - or C^2 - surface depending on its actual degree of continuity.

Table 1: Comparison of intersections of a C^1 - surface with a C^2 - surface.

CAx system, tolerance	#arcs	#ctrlp	deg	cont	maxdist1	maxdist2
System A, tol 0.02	19	22	3	2		
tol 0.001	42	45	3	2		
System B, tol 0.02	13	54	5	0	0.0015	0.0015
tol 0.001	14	65	5	0	0.00029	0.0007
System C, tol 0.02	25	28	3	2	0.004	0.002
tol 0.001	103	106	3	2	0.00018	0.00015
System D, tol 0.02	109	329	4	1	0.00001	0.000001
tol 0.001	109	329	4	1	0.00001	0.000001
Our method, tol 0.02	19	22	3	2	9.6e-5	0.0011
tol 0.001	34	37	3	2	4.9e-5	0.00015
tol 0.0001	55	58	3	2	3.1e-6	1.2e-5

Table 2: Comparison of intersections of a C^2 - surface with a C^2 - surface.

CAx system	#arcs	#ctrlps	deg	cont	maxdist1	maxdist2
System A, tol 0.02	19	22	3	2		
tol 0.001	41	44	3	2		
tol 0.0001	68	71	3	2		
tol 0.00001	140	143	3	2		
System C, tol 0.02	25	28	3	2	3.8e-3	1.8e-3
tol 0.001	115	118	3	2	1.1e-4	1.5e-4
tol 0.0001	427	430	3	2	8.5e-6	1.6e-5
tol 0.00001	1205	1208	3	2	6.0e-7	1.1e-6
Our non-opt, tol 0.02	24	27	3	2	3.8e-3	1.9e-3
tol 0.001	41	44	3	2	2.3e-4	1.4e-4
tol 0.0001	66	69	3	2	5.3e-5	1.7e-5
tol 0.00001	108	111	3	2	4.0e-6	1.8e-6
Our near-opt, tol 0.02	20	23	3	2	6.5e-3	1.7e-2
tol 0.001	31	34	3	2	9.8e-4	5.7e-4
tol 0.0001	56	59	3	2	6.2e-5	3.1e-5
tol 0.00001	91	94	3	2	4.9e-6	4.1e-6

Table 3 : Comparison of intersections of a C^0 - surface with a C^2 - surface.

CAx system		#arcs	#ctrlps	deg	cont
System A,	tol 0.02	30	33	3	2
	tol 0.001	61	64	3	2
	tol 0.0001	111	114	3	2
	tol 0.00001	181	184	3	2
System C,	tol 0.02	24+12=36	40	3	2, 0, 2
	tol 0.001	96+22=118	122	3	2, 0, 2
	tol 0.0001	376+66=442	446	3	2, 0, 2
	tol 0.00001	1064+174=1238	1242	3	2, 0, 2

System C creates two C^2 curves, joined with C^0 continuity at the joint, indicated by 2, 0, 2 in the column for continuity in the table. In the column for number of arcs, for this system we give the number of arcs for the first curve plus the number of arcs for the second curve, as well as their sum.

From the tables, it turns out very clearly that the task to represent intersection curves is treated very differently in the systems!

- In some systems, decreased tolerance leads to an explosion of number of arcs, in others not. For fine tolerances, one can improve a bit for all systems.

- Some systems represent the intersection curve with a C^2 curve though the true intersection curve is only C^0, others return a C^0 curve in this case.

- Some systems represent the intersection curve with a curve that is only C^0 or C^1 though the true intersection curve is C^2, other handle it well!

We have noticed that CAD quality managers, being aware on the mean effect that a coarse tolerance can cause in later stages of the design process, e.g. on conversion from one system to another, orders the finest tolerance possible in the earliest stages of design. From this point of view, it is extremely important, among others, that the systems utilise methods to approximate intersection curves that keep down the complexity of the represented curve, say, measured by the number of control points.

Concerning smoothness of the curve representing the intersection curve, to keep geometric faithfulness, the representation should mimic that of the true intersection curve. While, for some applications, it would be convenient to use a smoother representation, apart from not modelling the actual situation quite correctly, such a representation may generate undesirable features, like extremely small radii of curvature and more segments than really needed.

Using a less smooth representation than needed seems even more inappropriate, even if it clearly decreases the number of segments needed. As an example, using a C^1 curve of degree 3 instead of a C^2 curve of degree 3 gives 11, 26, 41 and 69 arcs instead of the 20, 31, 56 and 91 for our near-optimal method reported in Table 2.

5 Our representation method and its theoretic basis

A computationally inexpensive method to find a cubic spline approximating a smooth function is to interpolate the function at a number of points. More precisely, if f is a continuous function defined in an interval $[a,b]$ and $a = x_0 < x_1 < x_2 < \ldots < x_{n-1} < x_n = b$ are $n+1$ distinct points in the interval, then we search a cubic spline s such that

$$s(x_i) = f(x_i) \text{ for } i = 0, 1, 2, \ldots, n.$$

To get a unique solution, one needs to add further conditions, like for instance

$$s'(a) = f'(a), \quad s'(b) = f'(b),$$

where prime denotes derivative. The spline is determined through the solution of a tri-diagonal system of linear equations, which requires less than $15n$ floating number operations. This is well known and likely efficiently implemented in every CAx system for years.

One natural way to assess how well the spline represents the function is to measure the maximum deviation between them. The classical answer to this question was given as early as 1976 by the following theorem, due to Hall and Meyer [2]:

Let $a = x_0 < x_1 < x_2 < \ldots < x_{n-1} < x_n = b$, $h = \min_{1 \leq i \leq n-1} |x_{i+1} - x_i|$ and $f \in C^4[a,b]$.

If s is the (unique) cubic spline for which $s(x_i) = f(x_i)$ for $0 \le i \le n$ and

$$s'(a) = f'(a), \quad s'(b) = f'(b) \text{ then}$$

$$\|f - s\|_\infty \le \frac{5}{384} \|f^{(4)}\|_\infty h^4.$$

Here, $C^4[a,b]$ denotes the space of all functions defined in the interval $[a,b]$ having a fourth order continuous derivative, $f^{(4)}$ denotes the derivative of order 4 and, for a continuous function g defined in $[a,b]$, $\|g\|_\infty = \min_{a \le x \le b} |g(x)|$. Moreover, the constant $\frac{5}{384}$ is best possible.

In [5] one also finds similar estimates for quintic splines that may be used for cases in which one may want to represent an intersection curve with splines of degree 5. Since this application is entirely analogous to the way we will present below, we will not push it any further.

We would however mention that the technique used in [2] was recently reviewed by C. deBoor in [3], with the purpose to present the most general situation in which this technique is applicable, and to which we refer for the basic result needed for representations of intersection curves with splines of degrees other than 3 and 5 with our method. One should not increase much in degree however, since different organisations have a bound on what is acceptable, in many cases 5 and in any case we are aware of, below 10.

We also remind on the well-known result established in introductory courses on numerical analysis, see e.g. [4] that a similar estimate holds for cubic Hermite interpolation in which the first order derivative is also interpolated at the breakpoints, i.e. when $s'(x_i) = f'(x_i)$ for $0 \le i \le n$. If s denotes the cubic spline with knots of multiplicity two at x_i for $1 \le i \le n-1$ then $\|f - s\|_\infty \le \frac{1}{384} \|f^{(4)}\|_\infty h^4$. This kind of spline, which in general is only C^1, is used in our method to represent intersection curves when C^1 – continuity is desired.

Representation of Intersection Curves

The basic idea in our method is to use local variants of these results in the selection of interpolation points. The algorithm simply runs as follows:

1. Select a maximum allowed deviation d between the function f and the interpolating spline s.

2. Let $a = z_0 < z_1 < \ldots < z_n = b$ be a (fine) subdivision of the interval $[a,b]$.

3. Put $x_0 = z_0$.

4. Assuming that x_i has already been chosen and letting

$$M_{i,k} = \max_{i \leq j \leq k} | f^{(4)}(z_j) |$$

choose $x_{i+1} = z_{k-1}$ where k is the smallest integer satisfying

$$| z_k - x_i |^4 \cdot M_{i,k} \cdot r > 384 \cdot d$$

with $r = 1$ in case of cubic Hermite interpolation and $r = 5$ in case of cubic spline interpolation, provided that $k < n$, else $x_{i+1} = z_n$.

In this way we get a subdivision $a = x_0 < x_1 < \ldots < x_m = b$ of the interval $[a,b]$ such that when the function is interpolated by the spline at these points, the deviation between the function and the spline will not exceed d at any of the points z_i.

If the subdivision $a = z_0 < z_1 < \ldots < z_n = b$ is fine enough, one might assume that a maximum deviation of at most d is satisfied through the interval. In our application, this is not critical, since the maximum deviation of the curve to the two intersecting surfaces is always measured, giving the ultimate test.

Next we will use this method to improve a representation of intersection curves. Our point of departure is a representation of the intersection curve as a parametric curve. The requirements of the curve are:

- It must approximate the true intersection sufficiently well. If the improved representation is required to be within d from each of the two surfaces, it is convenient to assume that the parametric curve we depart from should be within $d/10$ or less from the two surfaces.

- Its parameter interval must be known, for all values in the parameter interval, one must be able to find the corresponding curve point, and for all but its knots, one must be able to find its derivatives of the first 4 orders.

We want to remark that it does not matter how complicated the actual representation is. Our experience from Open Cascade shows that the requirements are easily satisfied, indeed even if using $d/100$ instead of the $d/10$ indicated above, and we expect the same holds true, or may easily be enhanced to hold true, in any CAx system.

In a pre-processing step, we investigate the actual smoothness of the surfaces at the points where the curve we depart from intersects the isoparameter curves corresponding to u-knots and v-knots of the intersecting surfaces. Parameter values of the curve we are searching corresponding to these points will be needed knots and given a multiplicity compatible with the smoothness of the curve. In the examples where we compare our method with the ones of different CAx systems, however, we have refrained from this and instead adopted the multiplicity to attain the smoothness of the curves of the system we are studying.

Next, we make a rather fine subdivision of the parameter interval, compute the points of the start curve at these parameter values and use them to set up an accumulated chord length parameterisation for the curve we are searching. We also compute the derivatives of the first four orders at the original parameter values and use these to find the fourth order derivative with respect on arc length.

To find the improved representation, we simply run essentially the above algorithm with respect on the three coordinate functions of the curve, with the exception that we put a new knot as soon as one of the components deviates more than $d/\sqrt{3}$.

In practice, as turns out of the above tables, the method gives a curve with a reasonable number of components, located somewhat closer to the two surfaces than required. For instance, the $d/\sqrt{3}$- rule targets a "worst case" in which all the three components simultaneously deviate about equally much, while this is not often the case in practice. In the near optimal case, mentioned above, we capture a part of the discrepancy by iterating the process with d replaced by a few larger values.

6 References

[1] Sheldon Katz and Thomas W. Sederberg, "Genus of the intersection curve of two rational surface patches", Computer Aided Geometric Design, Vol5, No X, 1988, pp 253-258.

[2] C. A. Hall and W.W. Meyer, "Optimal error bounds for cubic spline interpolation", Journal of Approximation Theory, Vol 16, 1976, pp.105-122.

[3] Carl deBoor, "On the Meir/Sharma/Hall/Meyer analysis of the spline interpolation error", in: M. D. Buhmann and A. Iserles *(edt.):"Approximation Theory and Optimization", Cambridge, 1997.*

[4] Charles F. van Loan, "Introduction to Scientific Computing", Prentice Hall, New Jersey, 1997.

7 Authors

Roger Andersson, Chalmers University of Technology and Göteborg University,

Eklandagatan 86, SE-412 96 Göteborg,

rogand@math.chalmers.se

Bo Johansson, Chalmers University of Technology and Göteborg University,

Eklandagatan 86, SE-412 96 Göteborg,

bo@math.chalmers.se

Reverse Engineering of Objects Made of Algebraic Surfaces Patches

Bercovier, M. [1]; Luzon, M. [2]; Pavlov, E. [3]

1) The hebrew University of Jerusalem, Jerusalem
2) The hebrew University of Jerusalem, Jerusalem
3) The hebrew University of Jerusalem, Jerusalem

Abstract: Given a sampling process performed on $3D$ objects made of algebraic surface patches including sampling errors, a robust and probabilistic method for the detection of the algebraic surfaces is introduced. The complexity of the method is $O(k^{2*A} + k*n)$, where n is the number of points in the sample, k is an upper bound on the number of algebraic surfaces expected to be detected and A is the minimal number of points needed to determine an algebraic surface of the highest degree. For a scene containing only quadrics the complexity is only $O(k^{2*9})$. The method detects the surfaces with a small probability of error. The stages of the method are composed of creating hypotheses of the algebraic surfaces, embedding those hypotheses in hyperplanes, defining a deviation measure on the set of hyperplanes and clustering the hyperplanes formed using the deviation measure defined. The cost achieved by the method is better than the costs of most existing methods in most cases. The solution obtained is robust. The method does not assume an exact knowledge on the number of algebraic surfaces of the objects. Examples are given which consist of planar and quadratic surface patches.

1 Introduction

Many applications in CAD, Reverse Engineering and Computer Vision contain a sample of $3D$ objects (see [1], [2]). Such objects are made of bounded planar, algebraic or free form (parameterized) surface patches. The sample results in a measurement of the objects by a large number of $3D$ points with sampling errors. These errors are unknown but are bounded. The problem is to reconstruct the objects from the sample.

In [3] an algorithm, which accepts as an input a set of points on or near an unknown manifold, and produces as output an approximation by a simplicial surface, is described. The complexity of the algorithm is $O(n \log(n))$ where n denotes the number of points in the input set. [4] introduces a general framework for segmentation of objects given by 3-D scattered data. The cost of the method utilized is

$O(n(1 + neigh^3) + n\log(n))$ where n denotes the number of points in the scene and $neigh$ represents the maximal number of neighbors for every point.

[5] contains the process of generating CAD models from objects represented by $3D$ points clouds. It describes algorithms for reverse engineering conventional solid objects (objects bounded by simple analytic surfaces, swept surfaces and blends) and groups them into a global procedure. The procedure is complex and requires dense and accurate data.

Industrial objects are usually made of simple algebraic surfaces patches, moreover the cases of quadrics and planes are of great interest. Standard methods to solve this problem include different variants of least squares approximation ([6], [7], [8], [9]), random sampling ([10]) and Hough Transform ([11],[12],[13]).

For objects made of algebraic surface patches the detection process results most of the time in an approximation since the measurement of the objects contains points with errors. Approximation can be done by different least squares approximations using many distance functions (see [8]). The objects can be made of algebraic surfaces with self-intersections, branches and even unbounded, while the sample can be simple. Hence there exist methods which use only bounded surfaces (see [8]). Another possibility is to add to the scene a bounding box in order to bound the surfaces. For the case where the sample is complex one should also search for surfaces with self-intersections and branches.

Least squares approximation and random sampling do not always provide a robust solution, moreover for some of the methods even when the answer is satisfying the cost is relatively high ([6] is in $O(n^3)$, [10] is in $O(nk^4)$, where n is the number of points in the input scene, k is the number of original algebraic surfaces of an input degree d, and A is the minimal number of points needed to determine an algebraic surface of degree d).

The Hough Transform method (see [11], [12], [13]) depends on parameter settings, there exist cases in which it leads to incorrect results. It implies that the space of parameters is divided into a grid of size M^N, where N is the number of parameters of the space of algebraic surfaces and M is the number of units of each axis. M must be large enough to separate the patches. Most of the examples are restricted to $2D$ cases.

[14] suggests a method based on least squares fitting of algebraic spline surfaces, it simultaneously approximates points and associated normals, which are estimated from the data. The cost obtained is $O(n^2 + h^2)$, where n is the number of points, h is the number of unknown coefficients. The crucial step in the surface fitting process is the estimation of normals. Thus there should be no creases in the data, the distribution of the data should be not too irregular and the level of numerical noise should be small. Moreover the application is limited to a single surface.

In a previous work the problem of parameter detection of planar patches in an unorganised set of points in space, was solved in details (see [15]). We present here an abstract of the work for reverse engineering of objects made of algebraic surfaces patches, more details are found in [16].

The abstract is divided into four parts. The first part contains formulation of the problem. A robust probabilistic method of low complexity for the solution of the problem is introduced in the second part. The third part contains experimental results. Conclusions and future research are outlined in the fourth part.

2 Definition Of The Problem

Let $\{P^i\}_{i=1}^n$, $P^i = O^i + \overline{\varepsilon_i}$, where $\overline{\varepsilon_i}$ is an unknown error vector, be a set of points in R^3, suppose that there exist unique k (unknown) algebraic surfaces $S_1,...,S_k$ and k unknown index sets :

$\{A_j\}_{j=1}^k$, $|A_j| = i_j$,

with i_j being the greatest integer numbers which satisfy:

$\{\cup_{l \in A_j} O^l\} \in S_j, j = 1,...,k$

And suppose that $\forall l;\ 1 \leq l \leq n;\ \forall i,j;\ 1 \leq i \neq j \leq k$:

$O^l \notin S_i \cap S_j$.

The problem is to determine an approximation to the algebraic surfaces $S_1,...,S_k$.

3 A Robust And Efficient Probabilistic Method

3.1 Embedding in the space of hyperplane

The first step is derived from observing that working in the space of algebraic surfaces is quite difficult, hence define an embedding from the space of algebraic surfaces into the space of hyperplanes as follows:

For each algebraic surface, of a given degree d,

$$AS = \{(x,y,z) \mid \sum_{|\vec{i}|=0}^{d} a_{ijk} x^i y^j z^k = 0\}$$

where $\vec{i}$ is the index triplet $\vec{i} = (i,j,k), |\vec{i}| = i+j+k, i,j,k \geq 0$.

The number of coefficients of a general algebraic surface of degree d (excluding the free coefficient) is given by

$$A = (\frac{d}{6})(d^2 + 6d + 11) \qquad (1)$$

Define a mapping Λ of each AS into a hyperplane in R^A as follows:

$$\Lambda : \{\sum_{|\vec{i}|=0}^{d} a_{ijk} x^i y^j z^k = 0\} \rightarrow \{\sum_{l=1}^{A} b_l \Psi_l + a_{000} = 0\} \qquad (2)$$

where $a_{ijk} x^i y^j z^k, (i,j,k) \neq (0,0,0)$ are arranged in an increasing order of degrees $i+j+k = 1,...,i+j+k = d$, and their order inside terms of the same degree, is predetermined. $b_l, l = 1,..., A$ are independent variables replacing a_{ijk}, arranged in this order. $\Psi_l, l = 1,..., A$ are new independent variables replacing the monomials $x^i y^j z^k$, arranged in this order.

One can verify that Λ is an embedding of the space of algebraic surfaces of degree d in R^3 in the space of hyperplanes in R^A, where A is given by (1), A is also equal to the minimal number of points needed to determine an algebraic surface of a given degree d.

Using the embedding Λ the problem is transformed from the space of algebraic surfaces into the same problem on the space of hyperplanes.

For solving the problem we introduce a deviation measure on the set of hyperplanes, which is relatively robust to errors in the coordinates of the sample points.

3.2 Robust deviation measure on the set of hyperplanes

The deviation measure is defined in definition 1 (below).

Definition 1. Given a collection of hyperplanes $\{\pi_i\}_{i=1}^m$ in R^n, in an area of interest T, define the deviation measure between each different pair of them as an ordered triplet $\Phi_{(\theta, d_1, d_2)}(\pi_i, \pi_j), 1 \le i, j \le m$, where:

(d.1.1) θ is defined as the angle between the unit normals, in the same direction of π_i, π_j $(0 \le \theta \le \pi/2), 1 \le i \ne j \le m$;

(d.1.2) $d_1(\pi_i, \pi_j) = \min_{x,y} \{d(x, \pi_j), d(y, \pi_i) : x \in \pi_i \cap T, y \in \pi_j \cap T\}$;

(d.1.3) $d_2(\pi_i, \pi_j) = (d_1(\pi_i, \pi_j) - \overline{d_1})^2$, where the average, $\overline{d_1}$, is taken over all the "distances" between each pair of different hyperplanes as defined in the definition of $d_1(\pi_i, \pi_j)$.

This measure of "distance" is robust versus measurement errors. There can be many other measures of "distances" which are also robust versus measurement errors.

The measure of "distance" is vital to the problem of clustering hyperplanes in multidimensional space.

3.3 Overview of the method

Let $\max - d$ be an upper bound on the maximal degree of algebraic surfaces one wishes to detect.

Let d be the degree of algebraic surfaces to be detected in this iteration of the method.

Let k be an upper bound on the number of algebraic surfaces *expected* to be detected. When k is not known it can be iteratively increased till convergence of the results.

Assume that the sample points are not too close to algebraic surfaces intersections.

If the method is executed only with the maximal degree of algebraic surfaces one wishes to detect, the method may detect reducible algebraic surfaces as one algebraic surface, for example $z^2 - 9 = 0$ describes two planes $z - 3 = 0$ and $z + 3 = 0$. To prevent this one should divide the method to iterations with increased d values from $d = 1$ till $\max - d$, in the end of each iteration the detected surfaces are saved, a classification method is defined and applied on these surfaces, the points assigned to these surfaces by this classification method are removed from the input data, d is increased and a new iteration, operating on this value of d and on the input data left by the previous iteration, begins.

We introduce the randomization process by defining the events as selections of subsets consisting of exactly A (distinct) points. The sample space is defined as the space of all possible selections of A (distinct) points, there are

$\binom{n}{A}$ events, all events have equal probability (the probability distribution is uniform).

Consider the space of all possible algebraic surfaces defined by A (distinct) points, there are two categories of algebraic surfaces in that space.

Definition 2. *A "pure"* algebraic surface is an algebraic surface formed by A points belonging to one of the original algebraic surfaces.

Definition 3. *A "mixed"* algebraic surface is an algebraic surface formed by A points belonging to more than one original algebraic surface.

The following method is detailed:

Stage 1, Randomization:

(1.1) Given the points $\{P^i\}_{i=1}^n$, choose at random according to the uniform distribution, c subsets consisting of exactly A (distinct) points (c will be determined in the sequel).

(1.2) For each subset compute the algebraic surface it defines. If the algebraic surface obtained is degenerate, discard the subset and choose another subset. The algebraic surfaces computed form the set of hypotheses.

(1.3) Embed the algebraic surfaces obtained as hypotheses, in hyperplanes by the embedding Λ defined by (2).

The execution time of this stage is $O(c)$.

For each deviation measure's component (d.1.1)-(d.1.3) (will be denoted "distances"), do stage 2, stage 3.

Stage 2, "Distances" Calculation:

(2.1) Compute the "distance" between each pair of hyperplanes, by the component defined above. Build a complete (for each different pair of vertices there exist an edge connecting them) weighted undirected graph $G = (V, E)$, whose vertices are the hyperplanes, with edges – costs equal to the computed "distances".

(2.2) Perform a clustering on the graph G as follows: Find a min-cost spanning tree (MST) of the graph G (e.g. using Prim's algorithm – see [17] or any other book on algorithms for a complete discussion of MST). Compute the average cost of edges of the MST. Define two neighboring hyperplanes to be in the same cluster if the "distance" between them is less or equal to a value defined by a function of this average.

Examine the clusters formed:

Definition 4. A *bridge* is a vertex in a min-cost spanning tree (MST) of the graph G, connecting clusters while not belonging to any of these clusters.

Bridges are not wanted in the result of our clustering because they represent hyperplanes (algebraic surfaces) which are far from the original hyperplanes (algebraic surfaces), after choosing representatives from the final clusters to serve as an approximation to the original algebraic surfaces, if *bridges* are also included one can get algebraic surfaces which are far from the original algebraic surfaces and the approximation achieved is poor.

Bridges are discarded by repeating the stage of computing the average cost of edges in the MST, but with excluding the edges with large costs from this average computation. Complete, as explained above, by defining two neighboring hyperplanes to be in the same cluster if the "distance" between them is less or equal to a value defined by a function of this average. This way the clustering process output new results, *bridges* aggregate in tiny clusters to be removed later, "good" clusters are clusters of a sufficiently large size.

Note 1. The deviation measure of "distance" causes separating *"mixed"* algebraic surfaces by putting them in small clusters and clustering together *"pure"* algebraic surfaces, since *"pure"* algebraic surfaces are typically closer to each other than *"mixed"* algebraic surfaces or *"pure"* and *"mixed"* ones, with close meaning by this deviation measure. There exist cases in which *"mixed"* algebraic surfaces are close to each other or close to *"pure"* algebraic surfaces and both cases contain even closer than *"pure"* ones, in these cases they will be clustered together or clustered with *"pure"* ones – but this does not impose a problem since in these cases they define a useful approximation to the original algebraic surfaces.

The execution time of this stage is $O(c^2)$.

Stage 3, Clusters Filtering:

(3.1) Retain only clusters of *"pure"* algebraic surfaces and discard clusters consisting of *"mixed"* algebraic surfaces, by selecting all clusters of size larger than a percentage of the maximal cluster size, or some other statistical measure defined on the cluster size, or a constant value.

(3.2) Filter each cluster set and retain inside it a fixed number of vertices. Retain the vertices with most minimal sums of "distances" from all other vertices in the same cluster set, where the "distances" are defined by a predefined one of the deviation measure's component (d.1.1)-(d.1.3).

The execution time of this stage is $O(c^2)$.

Stage 4, Choosing Representatives:

(4.1) Choose as representative of each cluster its barycentric center in the space of hyperplanes (convex combination of the hyperplanes equations with equal weights). The representatives constructed by this method are well defined if the hyperplanes are normalized to have the same free coefficient. Instead of the barycentric center, any other acceptable criterion (i.e least square approximation variants) can be used.

(4.2) Map the hyperplanes accepted as representatives back to the respective algebraic surfaces, by the inverse mapping of Λ. Name the algebraic surface obtained from the representative of cluster i: S_i.

The execution time of this stage is $O(c)$.

Thus the total cost of each iteration of the method is $O(c^2)$.

Next we define the classification method used in the end of each iteration:

- Embed each algebraic surface in a hyperplane by the embedding Λ defined in (2).
- Map each point $P^i = (X_i, Y_i, Z_i)$, $i = 1,...,n$ to the A dimensional point P_A^i by the following mapping :

$$(X_i, Y_i, Z_i) \rightarrow (X_i^1 Y_i^0 Z_i^0, ..., X_i^j Y_i^k Z_i^l, ...), \ (j+k+l) = 1,...,d.$$

The order of the arrangement of the coordinates of each point in R^A is identical to the order of the arrangement of the coefficients of each algebraic surface used in the embedding's definition.

- Assign a data point to an algebraic surface (hyperplane) if the distance between them in R^A, as distance between a point and a hyperplane, is less than a predefined tolerance.

$$P^i \in S_j \Leftrightarrow \{Dis\tan ce(P_A^i, S_j) \leq Err\}$$

Note 2. The number of subsets selected, c, can be determined by the expectation value of the corresponding binomial distribution (see [18]), if one chooses a positive constant expectation value E, then uses the upper bound on the total number of algebraic surfaces *expected* to be detected, suggested, k, then the number of subsets to be selected, is typically $O(k^A)$. In the experiments $E = 25$ was chosen, thus the number of subsets was $c = 25k^A$. The complexity of the method was then $O(k^{2*A} + k*n)$, where A is the minimal number of points needed to determine an algebraic surface of the maximal degree chosen.

For extracting algebraic surfaces of a single degree d, the complexity is only $O(k^{2*A})$, where A is as above.

For quadrics with full dimensionality we obtain $O(k^{2*9})$.

4 Experimental Results

A probabilistic algorithm based on this method was prototyped in MATLAB, and tested on several examples.

4.1 Examples

(1) Sampling points on a hemisphere on top of a $3D$ cube. The center of the cube was: (60, -20,30) and the edge length was: 200. The center of the hemisphere was: (60, -20,130) and the radius was: 80. The points were sampled from an internal grid of the boundary of the object, each point was perturbed with random uniform error in the range $[-10^{-3}, 10^{-3}]$:

The quadrics equations of the object were:
- $X = -40$
- $X = 160$
- $Y = -120$
- $Y = 80$
- $Z = -70$
- $Z = 130$
- $-120X + 40Y - 260Z + X^2 + Y^2 + Z^2 = -14500$

The algorithm was executed for $d = 1$ and $d = 2$. For $d = 1$, a lower bound on the probability of a point belonging to a plane facet was $p = 760/8760$, hence the algorithm was executed with $k = 12$. The following surfaces were detected (see Figure 1):

- $X - 0.0000Y - 0.0000Z = -39.9999$
- $X - 0.0000Y + 0.0000Z = 160.0002$
- $0.0000X + Y + 0.0000Z = -120.0000$
- $0.0000X + Y + 0.0000Z = 80.0004$
- $-0.0000X - 0.0000Y + Z = -70.0002$
- $0.0000X - 0.0000Y + Z = 130.0005$

The classification method was executed and the points assigned to these surfaces were removed from the scene. The algorithm was run on the data points left with $d = 2$ and $k = 1$, the following surface was detected (see Figure 1):

- $-120.0000X + 39.9982Y - 260.0174Z + 1.0000X^2 + 0.9999Y^2 + 1.0000Z^2 = -14502.3000$

The algorithm was activated 20 times on this example, each activation was with different randomization of points from the scene. 19 of the cases were completed with success, that means the algorithm detected the cube's facets and the hemisphere. On one single case the algorithm did not randomize a large enough number of triplets of points from the upper plane of the cube (it randomized only 8 triplets), hence the plane $Z = 130$ was not detected and only 5 planes and one hemisphere were reconstructed. This case is included in the cases of the method's failure. Such cases exist since the method is probabilistic. To lower the method's probability of failure one has to repeat the method several times, or to provide a larger upper bound on the number of surfaces *expected* to be detected, to the method.

To test the robustness of the method, we performed on this example 20 translations of order of the size of the object and 20 three-dimensional random rotations with different solid angles from the range $[0,2\pi] \times [0,2\pi] \times [0,2\pi]$ were applied. The algorithm was tested on the accepted scene and proved to be robust. In all cases 6 planes (the translated and rotated cube's planes) and one half of an ellipsoid (the translated and rotated hemisphere) were detected.

Figure 1 was produced using POV-ray software version 3.5 (see [19]).

Figure 1: The detected half of ellipsoid on top of a 3D cube – Example 1

5 Conclusions And Future Research

A probabilistic method for the detection of algebraic surfaces constructing objects was introduced. The method is essential in the reverse engineering process. It solves problems that are considered hard to solve and is faster than other known methods in most cases. It is robust and does not assume an exact knowledge on the number of algebraic surfaces of the objects.

The input sample can be sparse or dense while other methods assume that the sampled data is dense.

The method was tested on several examples and proved to be effective. However there exist cases in which it reconstructs small amounts of spurious surfaces in addition to the original surfaces. These surfaces can be discarded by different postprocessing methods. One such method is to repeat the process several times and take the intersection of the results, or use clustering methods to accept the original surfaces.

Note that when the error in the data is very large with respect to the objects dimensions, the clustering could lead to "impossible" surfaces such as complex quadrics. In that case the method would not identify some of the algebraic surface patches.

The authors are currently working on the extension of the method.

6 Acknowledgements

The authors wish to thank to M. Werman, L. Joskowitcz, D. Lischinsky and M.A. Perles for fruitful discussions and encouragement.

7 References

[1] K.A. Ingle, "Reverse Engineering", McGraw-Hill, New York, 1994
[2] T. Várady, R.R Martin and J.Cox, "Reverse engineering of geometric models – an introduction", Computer aided design, 29., 4., 1997, pp. 255-268
[3] H. Hoppe, T. Derose, T. Duchamp, J. Mcdonald and W. Stuetzle, "Surface reconstruction from unorganised points", Computer graphics, 26., 2., 1992, pp. 71-78
[4] R. Chaine, S. Bouakaz and D.Vandorpe, "A Graph-Anistropic approach to 3-D data segmentation", Proceedings of the international symposium on computer graphics, image processing and vision SIBGRAPHI'98, IEEE computer society press, 1998, pp. 262-269
[5] P. Benkő, R.R. Martin and T. Várady, "Algorithms for reverse engineering boundary representation models", Computer aided design, 33., 11., 2001, pp. 839-851
[6] Charles V. Stewart, "Robust parameter estimation on computer vision", Society for industrial and applied mathematics SIAM, 41., 3., pp.513-537
[7] G. Taubin, "Estimation of planar curves, surfaces, and nonplanar space curves defined by implicit equations with applications to edge and range image segmentation", IEEE transactions on pattern analysis and machine intelligence", 13., 11., 1991, pp. 1115-1138
[8] G. Taubin, F. Cukierman, S. Sullivan, J.Ponce and D.J Kriegman, "Parameterized families of polynomials for bounded algebraic curve and surface fitting", IEEE transactions on pattern analysis and machine intelligence", 16., 3., 1994, pp. 287-303
[9] P. Roosseeuw, "Least median of squares regression", Journal of the American statistical association", 58., 1993, pp. 1-22
[10] M. Fischler & R. Bolles, "Random sample consensus: a paradigm for model fitting", CACM, 24., 5., 1981, pp. 381-395
[11] J. Illingworth and J. Kittler, "A survey of the hough transform", Comput. vision graphics image process, 44., 1988, pp. 87-116
[12] N. Kiryati, Y. Eldar and A.M. Bruckstein, "A probabilistic hough transform", The journal of pattern recognition society, 24., 4., 1991, pp. 303-316
[13] R.O. Duda and P.E Hart, "Use of the hough transformation to detect lines and curves in pictures", CACM, 15., 1., 1972, pp. 11-15
[14] Bert Jüttler and Alf Felis, "Least-squares fitting of algebraic spline surfaces", Advances in computational mathematics, 17., 1.-2., 2002, pp. 135-152
[15] Michel Bercovier, Moshe Luzon, Elan Pavlov, "Detecting planar patches in an unorganised set of points in space", Advances in computational mathematics, 17., 1.-2., 2002, pp. 153-166
[16] Michel Bercovier, Moshe Luzon, Elan Pavlov, "Reverse engineering from exact and noisy data of objects defined by algebraic surface patches", Technical reports of the Leibniz Center 2003, [2003-10], 2003

[17] Corman, Rivest, Lieserson, "Introduction to Algorithms", The MIT Press Cambridge, Massachussetts London, England, McGraw-Hill Book Company New York St. Louis San Francisco Montreal Toronto, 1990
[18] Feller, "An Introduction to Probability Theory and Its Applications", John Wiley & Sons 3rd edition 1, 1968
[19] "POV-Ray software version 3.5", http://www.povray.org/

8 Authors

Michel Bercovier, The Hebrew University of Jerusalem, Givat Ram, 91904, berco@cs.huji.ac.il

Moshe Luzon, The Hebrew University of Jerusalem, Givat Ram, 91904, moshelu@cs.huji.ac.il

Elan Pavlov, The Hebrew University of Jerusalem, Givat Ram, 91904, elan@cs.huji.ac.il

Part VI: *PLM*

From Systematic Innovation to PLM

Cugini U. [1]; Bordegoni M. [2];

1) Dipartimento di Meccanica, Milano
2) Dipartimento di Meccanica, Milano

Abstract: In order to improve and optimize the product development process, innovative systems are today used in the various phases of product development process. One of the main issues is the lack of formalized and validated procedures allowing the systematic introduction of these tools in the design process. That leads to a loss in process efficiency and performances. Therefore, it is of interest to study and test some possible solutions for integrating innovative tools such as PLM (Product Lifecycle Management) and KBE (Knowledge Base Engineering), with CAI (Computer Aided Innovation) systems and tools for topological optimization within the product development cycle. Besides, methods for analyzing and measuring product development process improvements are also necessary.

1 Introduction

The increasing demand for being competitive on the market has driven companies to drastically reduce product development cycles. At the same time, the growing of CAD/CAE and virtual prototyping systems we have seen in the last decade has deeply modified the approach to design: the possibility to test varying technical solutions maintaining low costs and time has increased the level of confidence with which designers can propose "extreme" solutions, and at the same time the preliminary phases of product development cycle have been cut down in favor of testing solutions reached following a "trial and error" approach rather than adapting a systematic innovation process.

The combination of these issues has led to an unstable situation for companies that do not hold the monopoly in a specific industrial sector. The study presented in Miller and Morris, 4th Generation R&D, J Wiley, 1999 shows that:

1. 10% only of North American companies have put on the market a new product in the last decade of XX century;

2. 90% of new products put on the market fail within four years from their appearance;

3. less than 1% of patents fully pay back the people who took on the investments;

4. 80% of successful innovation is produced by clients instead of being produced by producers.

In order to improve and optimize the process it is possible to think of introducing the use of new technologies or new approaches that are progressively available, and to set up product development environments that are collaborative and based on interoperability paradigms of data and applications. In particular, some methodologies and systems supporting the various phases of the product development process have been developed: a) methods and tools mainly used in the conceptual design phase supporting the *systematic transferring of innovative solutions* among different technical areas by means of an abstraction of the process; b) methods for *topological optimization* that propose optimal geometrical solutions using a generative approach; c) *knowledge-based* methods that support designers' activity through rules and knowledge re-use, and that reduce product development time without effecting its functionality, quality and security.

These methods are today used locally and are not integrated at all. This fact does not allow their optimum and effective use within the process. Nevertheless, in a concurrent engineering view it is required that the various phases of product design, optimization and engineering are integrated as far as possible, since the design and product development process is indeed a continuous iteration among these phases.

In such a context, it is interesting to study and test some possible solutions for integrating innovative tools such as PLM (Product Lifecycle Management) and KBE (Knowledge Base Engineering), with CAI (Computer Aided Innovation) systems and tools for topological optimization within the product development cycle. The major reason for doing that is the lack of formalized and validated procedures allowing the systematic introduction and use of these tools in the design process. The technological solutions available today within each of the mentioned areas are several. Therefore, it is also necessary to have appropriate and effective methodologies for selecting the most appropriate tools provided by the context of the process and the functionality required.

The paper describes some issues concerning the integration of the various innovative tools within the product development process, and presents a methodology developed for selecting the most appropriate technological solutions within a product development process that satisfy some given requirements.

2 Issues about methods and tools integration

Three main innovative methods and tools have recently appeared supporting some phases of the product development process: PLM/KBE, systems for topological optimization and systems supporting systematic innovation (Figure 1). PLM/KBE tools are the evolution of traditional CAD systems, and are used in the product development process for defining and describing *solutions* to design problems. The systems support feature-based and rule-based approaches, knowledge acquisition and re-use, detailed design, etc. and extend in some way the solutions domain [1].

The other two methods are more focused on stating and describing *problems*. They aim at supporting designers in finding innovative solutions to some very common questions, or in finding new solutions.

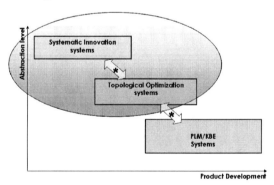

Figure 1: Integration of innovative tools

These methods operate at a high abstraction level: the design conceptual phase. In this area there are some methods and tools supporting systematic innovation [2]. They support the analysis of physical principles and system architectures that better answer to a set of given questions, and therefore that better solve a given problem. The solutions suggested by these tools are conceptual solutions that need to be described as geometrical elements that detail the shape of the system. Therefore, starting from the system architecture it is necessary to define the geometric envelop of the products, and the optimal shape and dimensioning of each component. Topological optimization methods and tools partly address this issue by providing an initial geometrical description [3]. These methods find optimal solutions that satisfy a set of input requirements and constraints. They usually start from an initial objective that requires to be optimized and generates a geometric solution that defines the shape of the product. Figure 2 shows an example of application of a topological optimization method where an optimized geometry and shape is computed starting from an initial design. When applying topology optimization almost all the information contained in the initial design is lost, since the optimized geometry is typically different from the original [4]. Topological optimization provides in output a geometry that does not take into account any standard, rule, manufacturing constraints, etc, and that requires subsequent fine tuning. A support to that is provided by feature-based and knowledge-based systems that allow a detailed definition of product geometry and shape that can be directly controlled and /or generated by rules, functions and procedures.

The integration objective of all the mentioned tools mainly concerns two main tasks:

1) make systematic the translation of the functional model of a system and of its design requirements into an optimization problem. That means identifying design variables, defining an objective function, defining design constraints;

2) define a Best Practice for the integrated use of topological optimization tools together with current PLM/KBE systems. That implies the definition of procedures for translating the topological optimization results into a geometry defined by "technological" features.

Both tasks answer to requirements that are today not yet satisfied in the product development processes, and those might bring advantages in terms of design time, cost and errors reduction, improvement in product quality, etc.

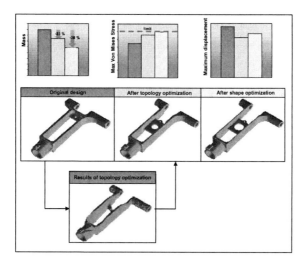

Figure 2: Example of topological optimization performed using Altair OptiStruct

3 Methodology for methods and tools selection

Two issues have to be considered for the adoption of new methods and tools within the product lifecycle: 1) how to select appropriate and effective methods and tools; and 2) how to estimate benefits and impacts before adopting and/or integrating those methods and tools. The authors have just concluded a research project proposing a solution to these two issues [5]. The aim of the project was to identify and validate an integrated environment for studying and evaluating the adoption of knowledge and innovation management tools within product lifecycle (www.kaemart.it/ike). The main outcome of the project is a methodology that consists of guidelines for the adoption and integration of new technologies within the product development process, including a method based on the QFD (Quality Functional Deployment) technique for candidate technology selection. The methodology is described in the following section.

3.1 Methodology

The methodology developed consists in a sequence of steps that should be followed for analysing a product development process of interest, and for selecting and evaluating the most appropriate technological solutions that might be integrated into the process with the aim of improving the performances of the process.

In order to evaluate the methodology developed we have carried out some test cases. One of them has been developed in collaboration with a company designing and manufacturing machine tools. This company's objective was to investigate the possibility of integrating KBE tools in their design process, to select the best tool(s) fitting within their already existing process and in-use technology, and to quantify and evaluate the benefits. The steps of the methodology developed are described referring to this specific test case.

The first step of the methodology consists in the selection and modelling of the product development sub-process as it is currently implemented in the company (As-Is process) by interviewing company's experts, collecting and structuring data. Process data are organized and structured using the IDEF0 technique [6]. The outcome of this activity consists of IDEF0 diagrams of the selected process, providing a well-structured view of the process, including activities, actors, roles, resources, technologies, information flow, etc. Process models are subsequently analysed in order to identify the most critical issues and those aspects that may be improved. In the machine tools test case, the company's product design process has been modelled and analysed, and information on knowledge has been extracted from it: types of knowledge, use of knowledge within the process lifecycle, how knowledge is stored, and its impact on the process, etc. This activity has allowed us to explicitly define activities related to knowledge in the process. This activity is typically carried out in collaboration with companies' experts (managers, designers, technicians) and process analysts, who help in structuring, rationalizing, and formalizing the process.

In parallel, some technology experts have updated the state of the art of emerging tools and systems aimed to knowledge and innovation management, also evaluating the level of implementation of each technology functionality and their integrability. This activity provides a useful overview of possible solutions for satisfying specific process requirements.

Then, the information produced in the analysis phase is collected into two matrices (Figure 3), according to the QFD (Quality Functional Deployment) method adopted [7]: *Matrix 1* that reports the importance of specific activities in respect to the process task of the selected product development cycle (in the test case, we report the importance of knowledge management activities in respect to each activity of the analysed design process); *Matrix 3* that links technologies with a set of functionalities related to the specific activities of interest (knowledge management activities in the reference

From Systematic Innovation to PLM

example). An intermediate matrix has been defined between Matrix 1 and Matrix 3 for correlating process activities and technology functionalities. The matrix, named *Matrix 2,* is filled in by process experts together with technology experts. In the test case, the matrix reports knowledge management activities along the rows and technology functionalities along the columns. The column "*valuation*" reports the relative importance values for each KM activity as resulting from Matrix 1. The importance of K-functionality related to each KM activity is set by the process and technology experts on the basis of how the process performances have to be improved (Bi,j values of Matrix 2). For each K-functionality we sum up the values measuring its relevance in respect to all EKM activities (weighted in respect to a specific process), and therefore, we obtain a ranking of the technology functionalities. These values are considered as weights in the "valuation" column of Matrix 3. Figure 3 shows the matrixes and the candidate technology selection process. The final selection of the candidate technology is done on the basis of the results obtained in Matrix 3. The decision is taken considering how a specific technology satisfies the product development process requirements.

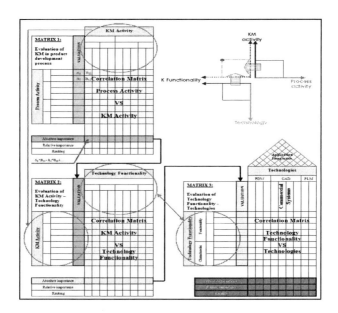

Figure 3: Candidate technology selection method

3.2 To-Be modelling and metrics

In order to evaluate the impact derived by the introduction of the selected technology in the product development process, a modelling of the To-Be process including the selected technology and a simulation of the new process are performed. Metrics parameters are defined in order to compare As-Is and To-Be processes, and are related to the specific product development process. Some metrics parameters are: time, errors, number of recycles, etc. At first, metrics permits us to refine the To-Be process, by evaluating performances of the To-Be simulated process versus the As-Is process; once the new technological solutions have been integrated within the selected candidate process, the metrics is applied to evaluate the process performance improvements, if any. If there are improvements in terms of quality and efficiency of the process, some specification of a new process based on the selected technological solutions are defined and subsequently implemented.

4 Conclusions

The paper has presented a view on an open issue that concerns the integration of innovative methods and tools within the process lifecycle: systematic innovation systems, topological optimization systems, and PLM tools. The paper has discussed the importance of integrating these methods for optimizing the process closely connecting conceptual design dealing with design problem statement within product detailed design dealing with design solutions. More research activities are required for fully addressing the issue of integration of these tools. One of the topics to address is the integration of various technologies and interoperability among them. The paper has also described a developed and evaluated methodology for selecting the most appropriate technology satisfying some product development process requirements.

5 Acknowledgement

The authors would like to thank MIUR (Italian Ministry for University and Research) for supporting the research, and M. Benassi, D. Pugliese, M. Pulli and M. Ugolotti from Politecnico di Milano, G. Cascini, M. Cambi and D. Russo from Univ. Firenze for their contribution to the research.

6 References

[1] Umberto Cugini, Michael Wozny eds., "From knowledge intensive CAD to knowledge intensive engineering", Kluwer Academic Publishes, 2002.
[2] Gaetano Cascini, N. Pieroni, Paolo Rissone, "Plastics design: integrating TRIZ creativity and semantic knowledge portals", Proc. of the Fourth International Symposium on Tools and Methods of Competitive Engineering, Wuhan, 2002.
[3] U. Schramm, "Recent Advances in the Application of Topology Optimization in Design", 2003.
[4] Pietro Cervellera, "Optimization driven design process: practical experience on structural components", ADM- AIAS Conference, 2004 (to appear).
[5] Monica Bordegoni, Gaetano Cascini, Stefano Filippi, Ferruccio Mandorli, "A methodology for evaluating the adoption of Knowledge and Innovation Management tools in a product development process", ASME International Design Engineering Technical Conferences & Computers and Information in Engineering Conference, Chicago (IL), 2003.
[6] IDEF0, Integration definition of functional modelling, Deaft Federal Information Processing Standard Publication 183 (FIPAPUB183). FIPA, USA, 1993.
[7] QFD, URL: www.qualisoft.com.

7 Authors

Umberto Cugini, Dipartimento di Meccanica, Politecnico di Milano,

Via La Masa 34, 20158 Milano,

umberto.cugini@polimi.it

Monica Bordegoni, Dipartimento di Meccanica, Politecnico di Milano,

Via La Masa 34, 20158 Milano,

monica.bordegoni@polimi.it

3D PLM Innovation through Openness

Philippe Laufer,

Dassault Systemes, Research & Development, CATIA Shape Design & Styling.

Abstract: This paper will describe the various criteria that should be evaluated to assess the openness of CAx/PLM systems, from openness to most commonly used standards, to high end frameworks provided to third party vendors to develop complementary solutions. It will be demonstrated that scalability in openness is essential to allow an ecosystem to grow efficiently around a PLM Solution like V5

1 Introduction

Deploying PLM as a major software vendor, following the strategic roadmap as identified in *Fig. 1* necessitates that product lifecycle development capability has significant breadth and depth. The specific needs of a company will further require that this capability be tailored, extended and integrated.

PLM is achieved through construction of the broadest and most integrated solution that is fundamentally based on the power of 3D to optimize the product lifecycle.

These requirements necessitate openness and the ability to reapply components to achieve different behavior or capability. The component application architecture provided within V5 has been specifically developed to meet these needs and is the cornerstone to build an ecosystem of software partners.

CAA V5 is Dassault Systemes open middleware and development environment for PLM. It delivers all necessary leading-edge technologies (PPR, Knowledge and many others) which are systematically used within Dassault Systemes PLM brands.

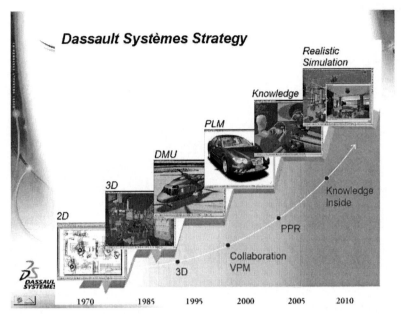

Figure 1: Dassault Systèmes Strategy

However, the openness of a system might not only be measured by the number of APIs it provides to C++ or Java Programmers; as PLM solutions should be scalable to fit each company's profile, ranging from Small and Medium enterprise to large OEM's, the openness paradigms should follow the same rule of scalability, for each company and each individual profiles.

V5 offers several level of tool sets and environment with the objective to propose to its CATIA, ENOVIA, DELMIA and SMARTEAM customers a large set of complementary and integrated applications developed, marketed and sold by members of its Software Community Program.

3D PLM Innovation through Openness

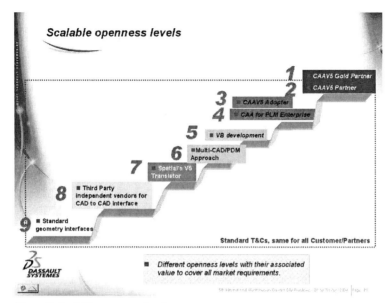

Figure 2: Scalable openness levels

2 Starting with openness to standards

As described in *Fig.2*, the openness of a PLM system starts with its ability to provide Data exchange interfaces using the most common standards (IGES, STEP, VDA and most recently announced like 3D PLM XML). This can be seen as the level one (or zero) of the openness represented by step 9 in *Fig.2* . However, the need often rises for customized translation and the ability for a PLM vendor to provide software components to third party vendors allowing them to develop customized Interfaces as if they were developed in house can be considered as the second level of openness marked as steps 8 and 7 in the step chart *Fig.2*.

DS, leveraging the long-time experience company in 3D technologies, Spatial , recognizes the need for efficient data exchange between native and non-native formats. Therefore, "3D InterOp" Translators are provided to allow developers to easily integrate advanced 3D data interoperability capabilities into software that utilizes 3D information, including CAD/CAM/CAE applications; data management systems, such as PDM, EDM, and MES; collaboration systems; and visualization solutions.

These InterOp Translators let gain access to 3D data without extensive code modification or purchase of expensive native software applications. It has revealed in many occasions that by incorporating such Translators into third party software vendor development strategy, has allowed them to focus more time on unique core competencies and value-add, reducing development time, lowering maintenance costs, and improving time-to-market

These components are aimed at transparent direct and indirect exchange of solid, surface, and wireframe data via a variety of neutral and native 3D formats, such as CATIA V5, CATIA V4, IGES, STEP, VDA-FS, Pro/ENGINEER (Pro/E), Parasolid (PS), Unigraphics (UG), SolidWorks, Inventor, and ACIS®. Each translator is fine-tuned and regularly updated to ensure accurate 3D data interoperability. The workbench for testing these translators are subjected to more than 225,000 tests weekly to identify areas for quality improvements

3 Openness to non expert Programmers

Passing beyond the need for pure interfaces or translators between heterogeneous systems, there is a strong needs for PLM customers to customize their system on site with complementary applications , in order to take into account specificities of their Process or in order to keep a competitive leading edge. As these industrials companies often do not employ high number of qualified software developers, the need for a simple tool to customize applications raised early in the CAD arena; Objective being for a User to be able to write his own macros to speed up specific processes. As an typical example, CATIA V4 had been providing for years a Module and a language called IUA (Interactive User application) that allowed any skilled user to build his own set of macros.

This way to approach customization for non developers has been standardized lately through Visual Basic and automation Concepts.

As an example , V5 allows to use a scripting language to access CAA V5 automation objects to capture own know-how and to increase productivity. The products that make up the CATIA and DELMIA applications share the same object model which can be accessed, as well as their own objects, by scripts written in Visual Basic with Windows, and scripts written in Basic Script for UNIX.

Scripts can be written from scratch, but can also use the journaling facility from the **Macros ...** command in the **Tools** menu that records end-user scenarios in scripts that can be then used as is or modified. This is definitively a jump start for occasional VB Script adopters.

An additional problematic that raises often is the interoperability of these tools for UNIX and Windows.

Typically, CATIA is an OLE Automation server for Windows NT and allows macro record and replay for both Windows NT and UNIX.

The macros recorded from the Tools menu and the Record Macro dialog box use the Visual Basic language, and not VBScript, to be compatible with Basic Script, allowing macro portability between UNIX and Windows NT, that is a macro recorded on Windows NT can be replayed on UNIX or the reverse. This means for example that Dim statements are recorded to declare objects as returned values of properties or methods.

As an example of how far this type of tools can be used, let's take the following example that can be automated using Scripting and Macros recording :
An important step in the process chain of the tooling process is the
calculation of the trimming edges and the analyses of the trimming edges
regarding the cutting angle. This step is performed during the development of the method plan after creation of the addenda surfaces on the design parts.

All tooling companies in the die industry need this functionality from the biggest car makers to small and medium companies
With CATIA Version 4 the functionality was fulfilled with a third party CAA Fortran written Application.

In less than a couple of weeks, an application specialist did a development on the basis of Macros and scripting, leveraging as well some knowledgeware interactive concepts to provide this capability to the users in V5; This is a clear example showing the progress of this Automation technology in the past years.

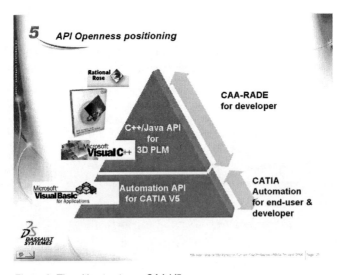

Figure 3: The ultimate steps : CAA V5

4 Rapid Application Development environement

As positioned in *Fig.3* above, the Component Application Architecture (CAA) V5 and its Rapid Application Development Environment (RADE) V5 provide developers with a platform for integrating in both Windows or UNIX environment, their expertise into specialized Product Lifecycle Management (PLM) applications, from initial Define through final product packaging.

3D PLM Innovation through Openness

The tools and Frameworks that are proposed in this set are equivalent to the one used internally by DS development teams :

- Developed applications are fully integrated in V5, in a durable way, taking benefits of V5 open-standard architecture. (see *Fig.4* for an overview of these standards)
- Methods and guidelines are proposed for building stable applications supported by a component-based architecture. (see *Fig.5* an overview of the current V5 Component based Architecture)
- A single build environment is included to generate run-time applications for both Windows and UNIX V5 applications. This means true cross-platform portability and reduced (to zero) time-to-port on several operating systems.
- The extended development CAA Configuration, targets C++ developers and allows full code implementation (including V5-specific wizards) and visual modeling capabilities for extending data objects from within CATIA V5 or ENOVIA V5. It also contains the full suite of test and quality tools, that DS has built through the past decades.

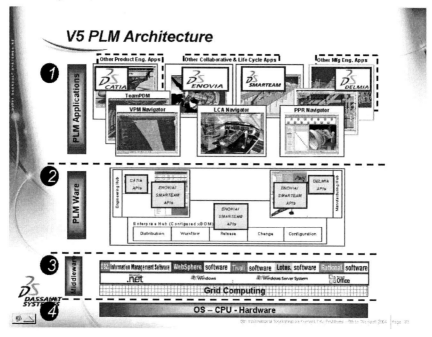

Figure 4: V5 PLM Architecture

Let us have a closer look at some of the Components/Collection of Development Tools that are typically necessary to develop PLM applications, and that are provided within the RADE (Rapid Application Development Environment) suite; it should clearly demonstrate that opening a PLM system is not to be reduced to provide a collection of APIs, but to provide a complete workbench allowing to 1- Design and Implement applications ; 2- Manage its quality; 3- Deliver and maintain the application

CAA - C++ Interactive Dashboard (CID)

CID is the single-point of access to all CAA RADE products, as well as the C++ development tools that support the full development cycle - from design and development through testing, deployment, and maintenance. It is tightly integrated with Microsoft Visual Studio C++® and Microsoft Visual C++ .NET 7.0 .While operating on Windows, CID allows automatic code building for UNIX platforms. Its main purpose is to Design and generate Wintop applications graphically, as well as Design and generate Application Logic.

Figure 4-1: view of CID CAA - Data Model Customizer (DMC)

DMC provides design-modeling tools for customizing or extending data object and features in CATIA V5 and ENOVIA V5. Using the same interface, you can create and extend objects from different CATIA modelers, while operating in a standard interactive graphic environment, based on industry-standard UML design modeling tools.

3D PLM Innovation through Openness

Access to DMC functions and commands are provided through the CAA - C++ Interactive Dashboard (CID) product.

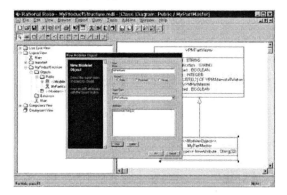

Figure 4-2: view of DMC CAA - C++ Unit Test Manager (CUT)

CUT facilitates test and quality control tasks critical to the efficient development of reliable software. CUT allows checking development compliance with design scenarios and ensure regression-free modifications and scenario pertinence, through features such as debug/non-debug replay, replay environment concatenation, etc.

Critical to ensure high quality and performance for end users applications, memory management is one of the outmost important task to be considered when developing C++ applications. Therefore, CUT also provides memory management test replay to help find memory leaks and ensure leak-free code and run-time error identification. During automatic test replays, CUT initiates coverage tests that guarantee that the entire code is validated

CAA - C++ Source Checker (CSC)

CSC aims to facilitate test and quality control tasks essential to the efficient development of stable software operating at the source stage in the application development cycle, CSC performs early checks against C++ coding rules to ensure better stability and reduce defects. This tool is typically the one that is daily used at DS to ensure that every single line of Code within V5 is following proper rules.

Figure 4-3: view of Source Checker CAA - C++ API Documentation Generator (CDG)

CDG automatically generates the C++ reference documentation of CAA-based applications. By integrating a single command into the release process driven by CAA - Teamwork Release Manager (TRM), CDG will generate C++ API reference documentation in HTML format. These HTML files - composed of the framework lists, interfaces, and classes list leading to the documentation page - are directly integrated into the V5 project development tree

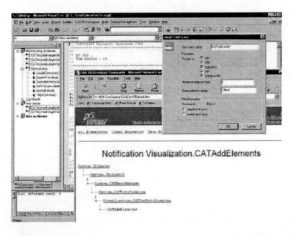

Figure 4-4: view of API Documentation generator CAA - Multi-Workspace Application Builder (MAB)

MAB delivers a consistent and integrated environment in which to compile, link-edit, and build a V5 application, using the same methods and tools that Dassault

Systèmes uses to create its V5 products. Industry standard compilers and linkers for languages such as C, C++, and Java are used with consistent processes and methods that are independent of the target platform. With its ability to handle multiple workspace compilation, link and run-time creation, MAB provides an efficient way to manage dependencies between separate workspaces

5 Conclusions

The subject of this paper was to explain the different ways a PLM system can be considered as open to plug in third party applications. Before looking at the panel of APIs provided, one should look at the scalability and complete set of tools that are provided to support concurrent development to the main vendor. Not only should the system be able to be customized by expert developers, but also by end users or casual programmers. The way APIs are provided and also organized reflects the Component architecture of the system (see hereunder *Fig.5* for V5 Component Architecture Overview). It is therefore a subtle way to understand how sound and lean the system is architected. So what rate would you give to the openness of the system you are using ? Answers can also be found looking at the size of the Software Ecosystem around PLM systems.

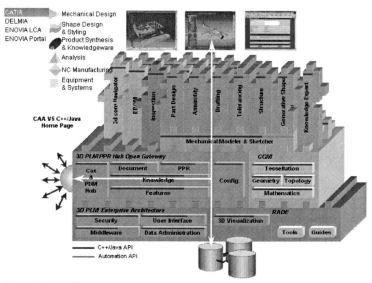

Figure 5: CAA V5

6 Author

Philippe Laufer, Dassault Systemes, Research & Development, CATIA Shape Design & Styling, Quai Marcel Dassault, Suresnes Cedex, BP 310, 92156 Frankreich, philippe_laufer@ds-fr.com

Part VII: *Strategic and Legal Aspects*

Embedding Niche Software into Heterogeneous CAD Environments

Salzmann, P.

ICEM Technologies GmbH, Hannover

Abstract: The last years have seen a consolidation of the CAD market. During the mid 90s the customers had the choice between several CAD packages (niche and platform packages). Nowadays the number of choices has been dramatically reduced. As seen in the office applications sector some years ago, concentration on a limited number of applications may lead to restrictions in the further speed of innovation and the reduced applicability of open standards.

This paper discusses this subject using an example of one specific area: The area of "Class A Surfacing", a small but nevertheless important niche in the digital design process of various products, for example in the automotive industry.

1 Introduction

A trend in past few years has been the continual reduction of innovation cycles. The ever increasing variety of automobile models and increasingly sophisticated designs of consumer goods are clear indicators of a revolutionary process triggered by rising customer demands and the manufacturers' competitive situation – a process which has not yet come to its end.

Customer demands are rising continuously, also with regard to the exterior design of products. Stylists, designers, and manufacturers need economical and efficient tools allowing them to meet these demands within competitive cost and time limits. Hardware and software solutions enabling a virtual product development are available today.

The "Class A Surfacing" which is used in this paper as example is the process stage in the digital design process delivering either the geometric components defining the appearance of the end product or the outer surface of technological parts as press or plastic tools. Figure 1 shows several examples.

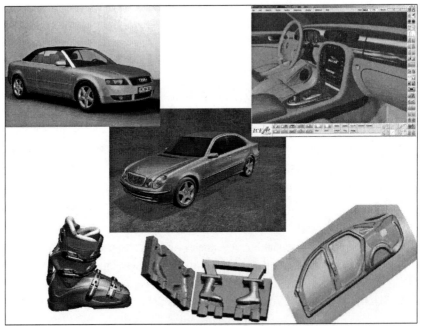

Figure 1: Typical Surfacing Products

2 The "Digital Design Process": a quick analysis

As in most application sectors the offered solutions differentiate from each other in terms of richness of functionality, efficiency and process compliance. In general it is possible to say that the available applications can be categorized as follows:

1. Specific functional applications embedded into platform CAD packages by the platform providers
2. Specific functional applications embedded into platform packages by specialized software providers using the infrastructure of the platform packages and
3. Highly specialized niche solutions.

Embedding Niche Software into Heterogeneous CAD Environments

Those offered solutions have several advantages and disadvantages, nevertheless the customers have the choice to find the balance between different approaches of the digital design process: The process compliance or the efficiency of tailored special solutions. So typical customer IT infrastructures differ significantly from complete monolithic environments working completely on one data representation to extremely heterogeneous environment containing several applications (see Figure 2).

Figure 2: Typical heterogeneous CAX environment

The arrival of parametric, history driven design approaches have influenced the views of the IT departments of CAX customers to focus more on data process compliance than to the functional richness of the applications (see [1]). This is supported by the fact that today there is no standard available which defines cross platform data structures containing more than geometry. And moreover there is no standard in sight, which may solve this problem in the near future (see [2]).

So the only choice seems to be to decide between proprietary data structures of suppliers, which have the long breath to be on the market in the foreseeable future. It may be worth to discuss the correctness of this approach.

2.1 Simplistic view of the digital design process

Since the author comes from the surfacing area, his experience is naturally based on the interplay between this process and the preliminary and following stages. But the problems and effects seen here can be applied to other stages as well.

A very simplistic (or sequential) view of the digital design process looks like this:

- The Styling stage, where it is more important to define the aesthetical appearance of the future product than dealing with the technical constraints
- The Surfacing stage, where styling concepts turn into the data description of the product used for manufacturing. Here the shape intent of the Styling stage is combined with the necessary fulfilment of technical constraints for the end product.
- The construction/tooling stage, where the surfaced components are combined with standard components (for a car: engine, power-train components, electrical equipment etc.) and where the connecting pieces are designed.
- The manufacturing stage, which starts from production process planning based on the complete digital product up to layout design of complete manufacturing sites.

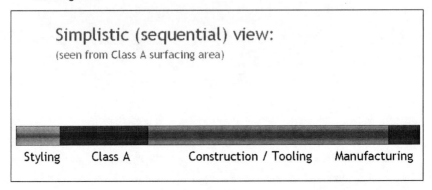

Figure 3: Digital Design Process

Of course in reality there will be some iterations turning in several executions of the process data flow in both directions. So in this sequential view it seems to be logical, that the efficiency in the data flow is impacted if there are gaps coming from different data representations and definitions leading to losses of special "knowledge" of specific process stages.

2.2 A more complex view of the digital design process

The reality today as a result of the "process compression" coming from the industry move to "Concurrent Engineering" during the late 90s the process today looks slightly more complex (see Figure 4).

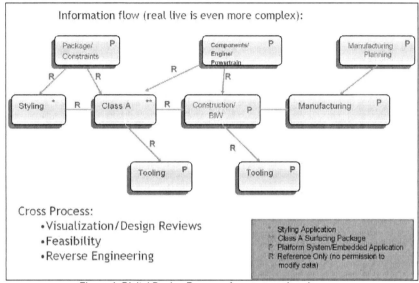

Figure 4: Digital Design Process: A more complex view.

This whole data flow is overlaid by tasks that have to be executed to enable verification of the current digital product state, as for example design reviews (performed by visualisation tools) or feasibility checks (performed by simulation tools).

In any product design process – whether it is a car or a consumer article, for example an Espresso machine, some constraints are given. So some standard components have to be used, for the design of a car it is clear from the beginning that a person has to fit in.

Those constraints are given as "package" into the design process. Normally there is already a data representation of the standard components available since they are a result of other design processes with very dedicated and specialised 'know how'.

It is therefore not allowed for the process stage referencing those "packages" to change the constraints directly. Normally any needed package changes, e.g. coming from another ("cooler"?) Styling, have to be required by the Styling department from the package provider.

For the data flow this brings, as logical consequence, a one-directional data stream, the following process stages are only allowed to reference the input data, and not to change them. This is not only true for our example of a new "cool" Styling requiring a new package - it is also valid for the communication between other process stages as Construction – Tooling, Construction – Manufacturing etc.

This also means that there is not necessarily any loss of specific "data intelligence" in this one-directional data flow because the process specific information always stays in the relevant area of competence.

This makes it questionable if it is truly necessary to have a monolithic, even proprietary data model for long-term storage of digital data, which belongs to the main assets of the customers using it.

In the following chapters some examples explain of how this requirement (the requirement to be compliant to a monolithic data structure) leads to unnecessary difficulties in technical, administrative and legal areas.

3 Challenges for a niche CAD system to fit into heterogeneous environments

3.1 Technical challenges

If a specialised niche CAD application has to be embedded into the complex IT environment as previously described, many prerequisites have to be fulfilled. Mainly the focus is set to data formats and structures. But in reality also the IT infrastructure with hardware (CPU, architecture, graphics cards), operating systems or the PLM environment belong to it.

3.1.1 Infrastructure

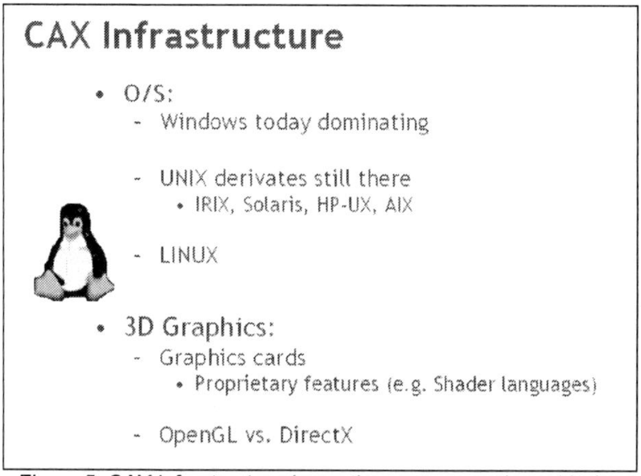

CAX Infrastructure

- O/S:
  - Windows today dominating
  - UNIX derivates still there
    - IRIX, Solaris, HP-UX, AIX
  - LINUX

- 3D Graphics:
  - Graphics cards
    - Proprietary features (e.g. Shader languages)
  - OpenGL vs. DirectX

Figure 5: CAX infrastructure (operating systems and graphics)

In the area of the operating systems today we see more and more the domination of one proprietary player who defined a de-facto standard in the office applications sector. As alternative there are Open Source approaches that are currently successful in the server area.

There are also other offerings, such as several UNIX derivates, but it is questionable how long they will be available.

In the future this may also impact the CAX application providers, because the move to one operating system provider will also influence the way in which CAX products are developed, for example in the 3D graphics area where today typically the open standard OpenGL is used as development platform.

3.1.2 Data Exchange

The exchange of geometry and simple attributes as structures, layers or colours has been solved for a long time - standards such as IGES or STEP and quasi-standards such as the STL format for facet data are available today and are robust enough to ensure a smooth bi-directional interplay between nearly all CAX applications. Problems coming from different topology or tolerance models of CAX packages remain, but with correct checking and refinement tools they are of minor importance and do not block the ability to deliver data to other process stages.

The need for design reviews requires more and more that graphical attributes such as material properties, textures, environments and lights are also used. Today's geometric standards like IGES or STEP do not fully allow the transfer of this kind of information. Approaches like Inventor or VRML are often used to transfer the required information.

Figure 5: Realtime visualisation of a CAD model containing several graphical attributes as lights, special material properties, textures and environment

But the real challenges today are parametric approaches and history.

History driven and parametric approaches have significantly contributed to make the digital design process more efficient, especially for iterative modification of digital data in the case of design changes. However a couple of problems exists with these approaches:

- Because there is no standardisation (moreover, it is questionable if there ever will be a standard which will support practically the loss-free transfer of history and parameters describing the digital data, see [2]) it is impossible to fully re-use the data in any other CAD application than the author system. Of course it is possible to transfer the geometry, but all "intelligence", such as feature information or dependencies in the history of the design, is lost. Even if it is possible to transfer the feature description, a history update in another system using different algorithms may lead to complete different geometrical representations of the product.

- There is still a contradiction between digital contents that can be described using parametric approaches (mainly parametric solids) and data following aesthetic design criteria (as Styling parts or free-form Surfacing parts). Especially designers and stylists feel restricted by the usage of parametric approaches. In addition all "automatic" changes coming from updating design history may lead to a "formal correct" result that nevertheless can be useless in reality because of the violation of aesthetic criteria which are impossible to describe (see also [3]).

To summarise, the full implementation of a homogeneous data flow using just one proprietary application seems to be the obvious solution. But this turns into a full dependency of the customer to the chosen solution provider, as it is extremely difficult to transfer the data using all the embedded intelligence to a new solution if it becomes necessary in the future. In reality the OEMs are changing their major CAD system all 8 – 10 years.

Coming back to the thesis described earlier that in reality only a one-directional data flow is really necessary there are approaches to overcome this by using a heterogeneous environment of different specialised applications. An integration level coming close to the usage of one proprietary application may be the usage of associativity between different applications, allowing each individual application to store their specific information where it has to be.

The following example describing the interplay between a Class A Surfacing solution and a platform CAD package demonstrates how this could be possible.

A free-form surface has been created using a specialised Class A Surfacing system using all the specific tools necessary to reach the required quality and aesthetic shape. The model is transferred then to a platform CAD package where it is referenced for construction work of underlying geometric structures.

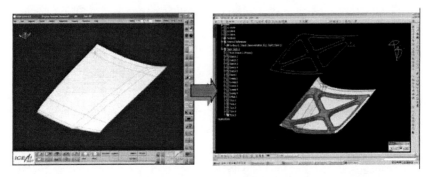

Figure 6: Transfer of Class A Surfaces to a CAD platform package, parametric add-ons in platform system

If the design is changed later, the part can be transferred again to the platform package. If the associativity is kept, there is no need to re-design the underlying geometrical structures, they can be simply updated. Of course this will only work well, if the changes are not too drastic. But bearing in mind the general restrictions of history updates the same would apply even if the design work would have been done purely in the platform package.

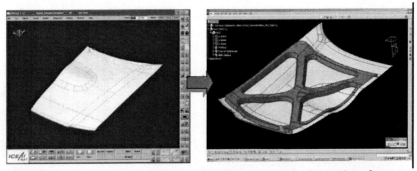

Figure 7: Update of Class A Surfaces in CAD platform package leads to update of parametric add-ons

This approach has many advantages for the CAX customer:

1. It is possible for each process stage to use the application that has the best specific tools fitting to it.
2. In a heterogeneous environment it is possible to replace or update just one tool, if the technological progress requires that. This is much easier than replacing a complete platform system.
3. The customer is not dependent to one solution supplier only.

So the data exchange standards should not focus to transfer features, they should offer solutions for transferring associativity.

3.1.3 Creation of embedded applications using APIs

Another possibility for delivering tailored solutions to customers who work in a homogeneous CAX environment is the usage of APIs (API: Application Programming Interface). With APIs it is typically possible, to embed own solutions and algorithms into a target CAD application as a 'plug-in'. Because the user interface and the database infrastructure of the host system is used, those solutions appear totally embedded and integrated.

While this sounds ideal in theory, there are several real challenges to bring this approach to life:

- The APIs are mainly commercial products, which have to be purchased by the solution provider. Sometimes it is not even possible to just buy them, an agreement has to be reached with the host system provider restricting the flexibility of the small solution provider to bring a general solution to the market.
- The solution provider needs to have installations of the host system available. If solutions for several host systems are to be created, this can be very expensive.
- For the usage of an API an exact match of the operating system, the build and API versions and even the relevant compilers are necessary. If working for several host systems this requires significant administrative efforts.

- The execution of the solution also requires the host system being available at the customer site. So the customer has the cost of both solutions: the platform and the plug-in solution.

All those restrictions would of course not be there, if all CAX providers would fully support open standards for data exchange.

3.2 Administrative challenges

Because of the non-availability of open standards in some areas, the trend in recent years forced the small CAX providers to deliver direct interfaces. Direct interfaces directly convert data from one CAD package into the representation of another package. So a niche system can directly import and export the native format of the chosen CAD platform package.

The idea was to avoid the intermediate mapping of data into a neutral data representation as currently required by standard interfaces.
In reality this idea does not work as expected:
Because every niche system has to be compliant to several platforms and cannot port its data representation to all partner systems, in real life the mapping of data using an intermediate data representation also takes place.
The amount of combinations arising from different operating system levels, hardware platforms, proprietary applications and independent release cycles makes it very difficult to administer.

Let's assume we have to support **8** direct interfaces on **6** Hardware platforms, supporting **2** O/S versions on each, on **2** release versions by interface. So we have:

8 * 6 * 2 * 2 = **192 combinations!!!**

This shows the complexity of the problem, especially if limited resources are available. If the same had to be achieved using open standards the number of combinations would be significantly less. Because of less formats, the standards are also more independent to the release cycles of CAX products.

3.3 Legal challenges

In addition to the technical and administrative challenges described above there are also legal aspects. Some examples:

- Software patents are protecting more and more technological claims. This is sometimes restricting the usage of very common quasi-standards (e.g. the GIF format).
- For co-operation, mutual agreements between solution providers have to be agreed. This goes from NDAs (non-disclosure agreements) up to the setup of complicated contracts dealing with fees, royalty splits.

Those legal challenges are cost and resource intensive, especially for small companies. Alone an ongoing patent research in all relevant areas can almost completely block small innovative companies.

4 Conclusions

We have discussed how small CAX providers have to deal with significant technical, administrative & legal efforts in order to bring their solutions to the market. As a result customers do not always receive the needed solutions in time – they do not receive the best possible, but under circumstances, only the best achievable solutions.

Therefore CAX customers need to think about their future requirements in terms of both functionality and process integrity, and should actively influence CAX providers. The goal should be to remove the need for dependence on only a few CAX solution providers.

Oppositely, the large CAX application providers should work more openly with customers and small, specialised providers. This is in the advantage of all parties:

- Small and large CAX providers
- And most important: **For the Customers!**

5 References

[1] Alfred Katzenbach, The fourth Workshop on Current CAx Problems, "Bridging the Gap – Features as the backbone for Integrated Engineering", Sanbi Printing Co. Ltd., Shizuoka, 2001
[2] Akihiko Ohtaka, The 5th Workshop on Current CAx Problems "Status of the STEP Parametric Project", Kloster Maria-Rosenberg, 2004
[3] Alain Masabo, The 5th Workshop on Current CAx Problems ,"Has Parametric a Future?", Kloster Maria-Rosenberg, 2004

6 Author

Peter Salzmann, ICEM Technologies GmbH, Küsterstr. 8, D-30519 Hannover, ps@icem.com

Interfaces as Essential Facility and EC Antitrust Law

Martin Stopper

Lehrstuhl für Zivil- und Wirtschaftsrecht, TU Kaiserslautern

Abstract: Interfaces can generate a dominant position if they are the clue to access a specific market. Under these circumstances it can lead to an abuse of a dominant position if the dominator is not willing to disclose the appendant interfaces. The legal practice enacted by the European Courts differentiates the first and the secondary markets to answer the question whether an undertaking abuses its dominant position by "keeping its business secrets".

1 Introduction

What we all learned about technical progresses and innovations in the field of interoperable processes leads to typical legal topics addressed to the Antitrust Lawyer. Especially in the field of innovations any technical progress leads to questions about an undertaking´s position in the relevant market. If you answer this question like "It has a dominant position, because it is that good" it leads probably to this legal question: "If it has a dominant position, probably it abuses this position."

The abuse of a dominant position is regulated in Article 82 of the EC Treaty.

Article 82 of the EC Treaty

Any abuse by one or more undertakings of a dominant position within the common market or in a substantial part of it shall be prohibited as incompatible with the common market insofar as it may affect trade between Member States.

Such abuse may, in particular, consist in:

(a) directly or indirectly imposing unfair purchase or selling prices or other unfair trading conditions;

(b) limiting production, markets or technical development to the prejudice of consumers;

(c) applying dissimilar conditions to equivalent transactions with other trading parties, thereby placing them at a competitive disadvantage;

(d) making the conclusion of contracts subject to acceptance by the other parties of supplementary obligations which, by their nature or according to commercial usage, have no connection with the subject of such contracts.

2 Legal and economical background

Addressed to Firms who are "above the market": What is a "dominant position" and when do you "abuse" this dominant position?

Efficient businesses are run with a view to conquering markets, to the point where they may establish very strong market positions indeed. Holding a dominant position is not wrong in itself. It is the result of the firm's own effectiveness. But if the firm exploits this power to stifle competition, this is an anti-competitive practice, which constitutes abuse. It is therefore the abuse of the dominant position which is wrong and which is prohibited by Article 82 of the EC Treaty.

A firm holds a dominant position if its economic power enables it to operate on the market without taking account of the reaction of its competitors or of intermediate or final consumers. In appraising the firm's economic power, the Commission takes account of its market share and also of other factors such as whether there are credible competitors, whether the firm has its own distribution network, whether it has favourable access to raw materials, etc.

This is the legal background in its general sense. Our interest is about a specific legal and moreover economical problem of this field. Due to the innovations in interoperable processes we stated here it leads to the following legal question: What is the consequence of having a dominant position in a specific software domain, if a undertaking is asking for the release or disclosure of the specific interface to serve its products on a secondary market. Or as worded in the microsoft case: What's to do, if a company is ordered to "disclose complete and accurate specifications for the protocols" necessary for the competitor's server products to be able to talk on an equal footing with windows PCs.

3 The Dilemma Situation

The preliminary question to handle the Antitrust Problem in an appropriate way is the right classification of the legal quality of an interface. But we want to cut this discussion off, because irrespective the fact if an interface is a subject of copyright law or just a business secret - following is undisputed: In both cases we talk about a form of private property struggling with Antitrust law which stands for public interest here specifically stands for unrestricted competition. To make it clear we talk about the dilemma between the interests of copyright law and Antitrust law.

The discussion about this conflict is brought together in the "Essential Facility- Doctrine". This doctrine was created in the U.S.-Antitrust-Law, specified in the trial United States v. Terminal RailRoad Association, 224 U.S. 383, 32 S. Ct. 507 (1912).

To take a classic example of the essential facility doctrine it is instructive to consider the case where access to a port is indispensable in order to be able to provide maritime transport services in a given geographical market. For the purposes of such a case it may be assumed that the owner of the port uses that infrastructure on an exclusive basis in order to secure a monopoly over the market for maritime transport services refusing without any objective justification to provide the necessary port services to arm's-length undertakings, which make a request in that regard. I consider that in such a case the case-law on the refusal to grant a licence must apply irrespective of the fact that the port services are not offered on the market. That fact does not preclude the possibility of identifying a market in port services requested by the maritime transport undertakings given that there is an actual demand for such services and there are no obstacles of a technical nature to the marketing thereof. In terms of the case-law on the refusal to grant a licence it may therefore be held that, by denying without justification access to the port infrastructure, the owner of that infrastructure would be abusing its dominant (monopoly) position on the market for port services inasmuch as by its conduct it would be eliminating any competition on the secondary market for maritime transport services.

4 Essential Facility in Case Law

Besides this classic example the European Court of Justice has given some important judgments concerning the Essential facility-doctrine. The specific legal questions in all these cases are following the same line of argument. The undertaking has a dominant position, their activities on the first market, their activities on secondary markets and tied to these alternative or cumulative activities: Do they abuse their dominant position?

4.1 The Magill Case

Giving judgment on an appeal from two judgments of the Court of First Instance in the well-known Magill case (ECOJ 1995 ECR I-743, C-241/91 and C-242/91) the Court had the opportunity of returning to the question of a refusal to grant a licence for the use of an intellectual property right. In the judgments appealed against the Court of First Instance had upheld a decision in which the Commission had adjudged that certain television broadcasters had abused the dominant position held by them on the market for their television programme listings, by invoking their copyright over such listings in order to prevent third parties from publishing complete weekly guides to the programmes of the various broadcasters.

In that connection the Court primarily emphasised that, although a refusal to grant a licence in respect of an intellectual property right cannot in itself constitute an abuse of a dominant position, exercise of an exclusive right by the proprietor may, in exceptional circumstances, involve abusive conduct. In that case, in the Court's view, the circumstances were such as to constitute abusive conduct on the part of the appellant broadcasters since:

- First, the appellants - who were, by force of circumstances, the only sources of the basic information on programme scheduling which is the indispensable raw material for compiling a weekly television guide - gave viewers wishing to obtain information on the choice of programmes for the week ahead no choice but to buy the weekly guides for each station and draw from each of them the information they needed to make comparisons. The refusal to provide basic information by relying on national copyright provisions thus prevented the appearance of a new product, a comprehensive weekly guide to television programmes, which the appellants did not offer and for which there was a potential consumer demand. Such refusal constituted an abuse under heading (b) of the second paragraph of Article 82 of the Treaty.

- Secondly, there was no justification for such refusal either in the activity of television broadcasting or in that of publishing television magazines.
- Thirdly, the appellants, by their conduct, reserved to themselves the secondary market of weekly television guides by excluding all competition in that market ... since they denied access to the basic information which is the raw material indispensable for the compilation of such a guide.

4.2 The Bronner Case

Then the Court had the opportunity of examining the problem of the refusal to grant a licence in the well-known Bronner judgment (ECOJ 1998 ECR I-779, C-7/97). In that case the Court, positing the existence of an autonomous market for nationwide home-delivery schemes, was required, inter alia, to assess whether the refusal by the owner of the only nationwide home-delivery scheme in the territory of a Member State, which uses that scheme to distribute its own daily newspapers, to allow the publisher of a rival daily newspaper access to it constitutes an abuse of a dominant position within the meaning of Article 86 of the Treaty, on the ground that such refusal deprives that competitor of a means of distribution judged essential for the sale of its newspaper.

4.3 The IMS Health Case

The latest and probably the most important judgment is the IMS Health vs. NDC Health case (CFI Case C 418/01).

Both parties to the proceedings are engaged in the collection, processing and interpretation of data concerning regional sales of pharmaceutical products in Germany. For present purposes, it is important to point out that the studies produced by those companies are structured on the basis of a geographical criterion under which the data on the sales of medicines are grouped together in a series of areas into which Germany is subdivided.

Those structures came into existence in response to various factors, such as the political boundaries of the municipalities and postcode areas. Detailed demarcation of segment boundaries is determined by other factors such as for example whether an urban or rural district is involved, communications and geographical concentration of pharmacies and doctors' practices.

IMS developed the so called 1860-brick structure which became the market leader. These IMS structures are data banks (or parts thereof) which are protected by the German copyright law.

Those structures were not used by IMS only for market reports sold to the pharmaceutical companies but were also distributed free of charge to pharmacy accounting centres and associations of health insurance schemes. Consequently, according to the matters mentioned by the referring court, those structures became a normal standard for the compilation of regional evaluations of the German pharmaceutical market. The pharmaceutical industry has adjusted its marketing and electronic data retrieval systems in line with them.

As a result of this marketing-strategy the market leader IMS was asked from their competitor NDC to grant to it for valuable consideration a licence to use its structure over 1860 areas. In response to the refusal by IMS to grant it such a licence NDC lodged a complaint of abuse of a dominant position with the Commission.

On basis of these matters the Commission therefore considered that the so called 1860-brick or compatible structure was indispensable to compete on the relevant market. Taking the view that there were no objective grounds for refusal of a licence the Commission accordingly held that such refusal constituted a prima facie abuse of a dominant position.

By applications lodged on 6 August 2001 IMS applied to the Court of First Instance for annulment under Article 230 EC of the Commission Decision and for suspension of operation under Article 243 EC. By order of 26 October 2001 the President of the Court of First Instance granted the application for interim suspension.

In the case brought by IMS before the Court of First Instance for annulment of the Commission Decision proceedings were suspended by order dated 26 September 2002 pending delivery of judgment by the Court in the present case.

On 2nd of October 2003 the opinion of general advocate Tizzano was published. It gave a new and differentiated analysis for dealing with the essential-facility doctrine This is the summary of his essential statements:

In view of the fact that that question proceeds on the assumption that the brick structure for which the licence was sought is essential to the marketing of the studies on regional sales of medicines in a given country, it is not hard to identify an upstream market for access to the brick structure (monopolised by the owner of the copyright) and a secondary downstream market for the sale of the studies.

That said, I must none the less add that the judgments of the Court on the refusal to grant a licence over an intellectual property right lead me to believe that, in order for

Interfaces as Essential Facility and EC Antitrust Law

an unjustified refusal to be deemed abusive, it is not sufficient that the intangible asset forming the subject-matter of the intellectual property right be essential for operating on a market and that therefore, by virtue of that refusal, the owner of the copyright may eliminate all competition on the secondary market.

Even where those circumstances obtain, in weighing the balance between the interest in protection of the intellectual property right and the economic freedom of its owner, on the one hand, and the interest in protection of free competition, on the other, the balance may in my view come down in favour of the latter interest only if the refusal to grant the licence prevents the development of the secondary market to the detriment of consumers. More specifically, I consider that the refusal to grant a licence may be deemed abusive only if the requesting undertaking does not wish to limit itself essentially to duplicating the goods or services already offered on the secondary market by the owner of the intellectual property right but intends to produce goods or services of a different nature which, although in competition with those of the owner of the right, answer specific consumer requirements not satisfied by existing goods or services.

That was in my view clearly held in the Magill judgment in which, as has been seen, the Court held an unjustified refusal to grant a licence to be abusive, inasmuch as (a) it prevented the appearance of a new product, a comprehensive weekly guide to television programmes, which the appellants did not offer and for which there was a potential consumer demand; and (b) by way of that refusal the appellants [had] reserved to themselves the secondary market of weekly television guides by excluding all competition in that market.

Therefore the advocate general considers that the reply to the first question should be that Article 82 EC must be interpreted as meaning that the refusal to grant a licence for the use of an intangible asset protected by copyright entails an abuse of a dominant position within the meaning of that provision where (a) there are no objective justifications for such refusal; (b) use of the intangible asset is essential for operating on a secondary market with the consequence that by way of such refusal the owner of the right would ultimately eliminate all competition on that market. However, that is subject to the condition that the undertaking seeking the licence does not wish to limit itself essentially to duplicating the goods or services already offered on the secondary market by the owner of the intellectual property right but intends to produce goods or services of a different nature which, although in competition with those of the owner of the right, answer specific consumer requirements not satisfied by existing goods or services.

The advocate general's opinion was confirmed by the Court of First Instance (Judgement from 29 April 2004/C-418/01) who stated: An abuse of a dominant position within the meaning of Article 82 EC is constituted where the following conditions are fulfilled (cumulative):

- the undertaking which requested the licence intends to offer, on the market for the supply of the data in question, new products or services not offered by the copyright owner and for which there is a potential consumer demand;
- the refusal is not justified by objective considerations;
- the refusal is such as to reserve to the copyright owner the market for the supply of data on sales of pharmaceutical products in the Member State concerned by eliminating all competition on that market.

5 Conclusion

If you have a dominant position on a market and you are not willing to serve an existing secondary market, you abuse your dominant position. If you have a dominant position on a market, which originates a secondary market where you are acting too, you only have to grant licenses to competitors who are capable to create different business concepts. The relevant indicator for this judgement is always the customers welfare.

5 Author

Dr. Martin Stopper, TU Kaiserslautern, Lehrstuhl für Zivil- und Wirtschaftsrecht,
Postfach 3049, 67653 Kaiserslautern,
stopper@rhrk.uni-kl.de

Outlook for the CAx/PLM Future

C. Werner Dankwort

University of Kaiserslautern, Germany

Abstract:

Since many years the industrial processes are supported by software tools, CAx and PLM are the actual abbreviations. The industrial globalisation with a very tough international market has impact on the interplay between product generation processes and the CAx/PLM world. For both fields significant changes have to be expected.This paper try give prognoses about the a long term future of CAx/PLM out of the industrial point of view. Speculations basing on today's innovation with long years cycle time for industrialisation will be given.

1 Introduction

In technical development – both for industrial products and for IT-systems – people often speak about "Strategies", "Outlook for the future", or "Long term trends". Serious motivations for such approaches may be: To avoid decisions with a negative long term impact on the market position and therefore to insure the survival of the enterprise – concerning both: Industry and software vendors.

Often the strategies follows simple extrapolations of the past or presence to the future. Some reasons, why it may be so difficulty to propose concepts for long term future, lies in the human thinking: It seems to be too risky to recommend strategic approaches, for which it can not be proved, that they can be realised, and – if it would be done – that the necessary investment will give a return in a limited time. The focus on short term economic arguments together with limited knowledge about the technical chances and visions will obstruct long term innovations.

There seems to be a lack within the organisational and cultural structure of enterprises: Visionary thinking is normally not a main issue of education of engineers and managers. But for the future with very fast developing techniques and their impact on the extremely fast changing industrial processes – and visa versa – also visions are required.

2 Today's Industrial Challenges in CAx

Considering the industrial situation today we may have various point of views (see Figure 1), concerning:

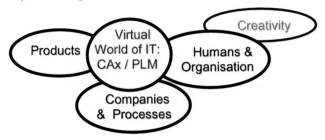

Figure 1: CAx/PLM in Industrial Environment

The **products**: That means to fulfil the requirements of the market, to satisfy the customers. Technical modern products like automobiles are the result of integration of mechanical engineering, materials with targeted developed properties, electronic components, microcomputer, software, etc. Nowadays software should be considered as an engineering part too.

Modern products will get more and more "intelligent", with self-learn capabilities, able to adopt themselves to the needs of the customer or to environmental conditions.

For CAx application the Product Information Model is of fundamental importance. Classically it contained mainly geometric information, later on also product attributes like materials and administrative data. But also process information has to be integrated in one common Product and Process Information Model. The focus on the product itself, that means the functionalities and properties from the customers point of view is still insufficient in the actual approaches.

The **companies**, cooperating today within a global network, do need very fast, safety information management and communication technology to fulfil the CAx/PLM requirements of global wide product development and manufacturing. The focus in the companies lies on the Product Generation Process (PGP) including the main parts of the life cycle: From the requirement of the market, via the product concept, the development until the production, sales, maintenance. Non-passive Product- and Process Information Models in a distributed management system, dynamically adapting to the progress of the processes, are still a challenge.

Outlook for the CAx/PLM Future

The **human potential** often is considered as most important for the success of a company together with its organisation, including all people within the companies from top management to the unskilled worker. The human resource management together with the "Personal Culture" are key factors for the image, the innovation power, the efficiency of a company and by that for the position of a company on the market. This field has to be discussed a little bit more.

3 Human Resources in the Industrial CAx-Future

The more and more intensive penetration by CAx of all industrial processes has a huge impact on the personal field. Classical jobs like draftsman or experts for releasing technical drawings are no more needed. But new job profiles are required [1] like:

- Coordination functions between the various tasks in the process
- Tasks for integrated product generation, combining engineering design, simulation, FEA, manufacturing estimation, controlling etc.
- Parametric-associative engineers
- Target oriented product development ("Product around the customer")

Besides the "New job profiles" there are a few other aspects concerning these topics:

- More and more complex CAx-systems may overstrain standard engineers
- Engineering skills are not mandatory coupled with CAx-skills
- Creativity is not supported by CAx systems (may be even restrained by them)

The human potential may follow something like a Gauss distribution (s. Figure 2).

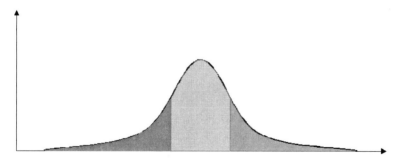

Figure 2: Human Potential and Gauss Distribution.

The complex, high sophisticated CAx-tools require people with high potential. There is no problem with people on the right – red – part of the distribution. But what about the majority in the middle? For them we have to be aware that these people may be not so skilled in CAx-technique, but possibly are very good engineers. To get out of this dilemma the companies see the solution in a very intensive, technically and psychologically elaborated training.

An additional alternative may be: Not to form the humans concerning the needs of systems, but to form the systems concerning the competence, the abilities, and the needs of the persons, which have to work within the product generation process.

An complementary aspect is the following: Not the system should develop the product, but the engineer, who is only supported by the "servant" CAx-system. That means, the product with the required properties and functionalities will dominate the digital representation and digital work and not the system specific CAx-technique.

4 CAx-support in the Past / Life Cycle of CAx- Innovations

Since the beginning of CAx there was always a "Three partner relation": Researcher (mainly Mathematicians – geometry oriented), data processing people (information technology, as it is called later), and people from application industry.

The impact of the last was over long time very small – although they were the customers. Therefore it took many years that the handling, the user interface of the systems get sufficiently. From all of the three partners came ideas, wishes, innovations, but that the three partners work together in really one team, was the rare exception.

Therefore it takes very long time until a new CAx-approach comes to broad industrial application. In Schedule 1 some examples are given.

Outlook for the CAx/PLM Future

	STEP	Reverse Engineering	Parametric	PDM	3D-CAD	Rapid Proto-typing
Innovative idea / First concepts	~ 1984	~ 1980	~ 1979/80	~ 1980	~ 1970	~ 1985
Projects Prototypes					~ 1978	~ 1988
Industrialisation First SW-Product	~ 1991 - 1994		~ 1986	~ 1985	1980	1992
Industrial applications in special areas	~ 1995	~ 1994	~ 1990	~ 1995	1985	1993
Area-wide industrial applications	~ 2000 +		~ 1994 (PTC) ~ 2002 (CATIA-V5)	~ 2000 +	1990	1995
Y e a r s	16 +	14	15 - 22	20 +	20 +	20

Schedule 1: Cycle Time of CAx-Innovations - Historically

Most of these information are well known, but only a few are aware how many years really run until there is a broad benefit of a new CAx-ideas. There are several reason for this facts, it would go out of scope of this paper to discuss them here. But looking into the future, we should judge the today's situation: Will the modern industry be able to afford further long years delay for new CAx-technology?

Looking to the challenges to industrial application mentioned above, the "three partners" have to find a way to decrease the industrialisation time to a few years.

5 Vision of Future CAx-support

Three objectives may be pitched out of industrial application experiences:

- Focussing the engineers thinking
- Product definition basing on customers point of view
- Supporting the complete product life cycle with the aid of integration of product information and process information

For optimal product generation process new kinds of CAx-software tools are required:

- The engineer will work directly on the required properties and functionalities of the product ("Property and function driven design")
- To control the product shape, the product structure, the engineering design, the engineer can use CAx-functions given by terms of the colloquial language between designers and engineers – or on long term – between designers/engineers and customers (leading to "Intent driven Design")
- To support the whole process, the best appropriate CAx tools from various vendors can be combined without any break within the information flow

These ideas go fare beyond the state of the art of today's CAx in industry, like parametric, associative design, and feature technique.

CAx support of product generation with these new tools will give a chance to achieve the objectives mentioned above. Is it realistic to think about this? Some first steps are already done successfully, which prove, that an long term we must be aware, that for future product generation with CAx we have to expect a paradigm shift. But does it mean again waiting more than 20 years?

6 Some Examples of Advanced CAx-Functionalities

In Schedule 2 some modern CAx innovations principally known are given. Considering the challenges in modern industry with the very fast changing processes, we must not wait so many years: The people involved can not change their mind with the same speed. Therefore the systems have to follow the thinking of the people and not vice versa.

Outlook for the CAx/PLM Future

	Open CAx-Component Architecture	Engineer. in Reverse	Optimis. Holistic Design	Knowl. Based Engin.	Product configuration	Design by terms of language	New active & intelligent engineering Elements
Innovative idea / First concepts	~ 1988	~ 1988	~ 1965	~ 1985	~ 1988	~ 1999	~ 2000
Projects Prototypes	~ 1995 - 1998	~ 1996 - 2001	~ 1970			~ 2003	
Industrialisation First SW-Product			~ 1975	~ 1990	~ 1995	? 2006 ?	
Industrial applications in special areas	~ 2000	? 2005 ?	~ 1975	~ 2002	~ 2000		
Area-wide industrial applications	? ?	? ? ?	? ?	? ? ?	? ?	? ? ?	
Y e a r s	16 +	16 +	22 +	(20 +?)	(17 +?)	? ? ? ?	???????

Schedule 2: Cycle Time of CAx-Innovations - Historically

A few approaches concerning new CAx-methodology will be sketched out:

Engineering in Reverse:

This concept was developed to design a product by working with the properties [2,3]. The principle is given in Figure 3. An application on surface modification was successfully implemented in the FIORES project.

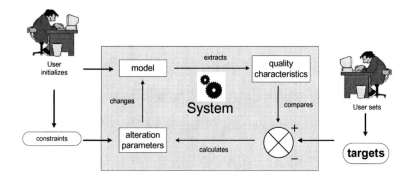

Figure 3: Engineering in Reverse: Holistic approach of Optimisation (FIORES [2,3])

Control of design by terms of language:

In the European project FIORES-II [4] this idea was persecuted for Aesthetic Design: The "Marketing Language" used by the final customer and the "Designer Language", used between stylists and designers were analysed with respect to terms used by various European Designers. The link between both languages was found by Artificial Intelligence methods (Case Based Reasoning), while the terms of the Designer Language were transferred to CAGD and could be used as modelling commands by the designers ("modifier") as given in Figure 4. An example is the term "acceleration" presented in Figure 5.

There is no reason not to extend these concepts to other engineering tasks.

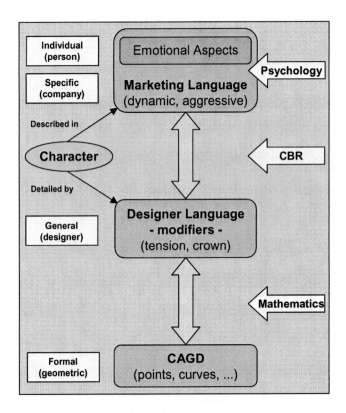

Figure 4: Design control by terms of languages (FIORES-II [4])

Outlook for the CAx/PLM Future

Intent driven design:

The results of the European project FIORES-II show, that it should be possible also to capture the "Design Intent", that mean subjective, non-rational evaluation of products. [4,5]

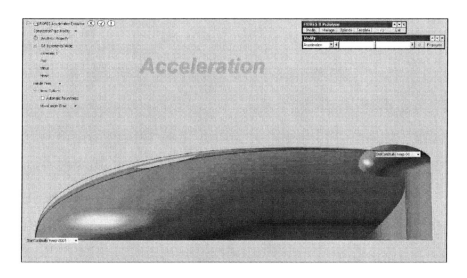

Figure 5: Term "acceleration" as a surface modifier (FIORES-II [4])

Full interoperability with open component systems:

Another vision mentioned above is the approach to use the best of class CAx-components together in the process – even from competing vendors – without problems in the information flow of the processes. Some basic concepts for a general interoperability are already available [6], but for special DMU application a system is running at a well know automotive manufacturer linking CATIA (V4) and Pro/Engineer [7].

To implement a general interoperability concept is more a question of the software market and the vendor situation then a basic technical problem.

7   Remarks about the Future CAx-Landscape

One approach to look into the future is to extrapolate the presence. The alternatives discussed here will also consider some visionary aspects sketched above without guarantee that there will be a chance for realisation within the next years:

Alternative A:

The CAx-Landscape will be dominated be one (or very few) system suppliers, with severe consequences for the application industry, but also for CAx-innovations and for the market of small 3rd party software vendors. The interoperability of CAx-components and CAx-PLM integration has to follow the vendors interests.

Alternative B:

The gap between "Midrange" systems and "Jumbo" systems will disappear. Another re-arrangement on the CAx-market will happen. Interoperability and integration CAx-PLM will be the challenge.

Alternative C:

Industrial processes will be the drivers of any CAx development. Some aspects may be discussed:

- Orientation of the system development on the industrial product itself, its properties and on its customers, orientation on the people within the processes
- The best CAx-tools on the market will be available, problems of interoperability must be solved, innovations have to be pushed, cycle times can be reduced
- Integration of CAx- and PLM tools will be possible

Pre-condition is a technical solution as open component architecture.

What really will happen can not be forecasted. Some predictions concerning system vendors we will dare:

The main business in the future will be process support and no more software development. The management structure and culture of modern industry with a network of suppliers will impact the management of software enterprises. The in-house development depth for software will decrease dramatically. The key of the future will be "system integration" and no more "functionalities". For software quality the philosophy of today of "Single Item Product" will adapt strategies from "Serial production systems".

8 Conclusions

We are aware, that any outlook into the long term future will have a wide field of speculations. But the future will not come following a superior law. It is in the responsibility of the persons today involved in the industrial processes to model the future to their need. But this premised that there is an understanding, what need and challenges of the future are. And that there is courage and cleverness to go this way. Here lies our common responsibility as the team of the three partners: Industry, software vendors and research.

9 References

[1] C. Werner Dankwort, Roland Weidlich, Birgit Guenther, Joerg E. Blaurock, "Engineers' CAx education – it's not only CAD", Computer Aided Design, Vol. 36, Issue 14, "CAD Education", eds.: Nickolas S. Sapidis, Myung-Soo Kim. page 1439-1450, 2004

[2] FIORES, "Formalisation and Integration of an Optimised Reverse Engineering Workflow" – Project of the European Community Brite Euram BE 96-3579, from 01/1997 – 12/1999, Coordinator RKK, University of Kaiserslautern.

[3] C. Werner Dankwort, Gerd Podehl: "A New Aesthetic Design Workflow – Results from the European Project FIORES", in P. Brunet, C. Hoffmann, D. Roller (Eds): "CAD Tools and Algorithms for Product Design", Springer-Verlag, 2000.

[4] FIORES-II, "Character Preservation and Modelling in Aesthetic and Engineering Design" Project of the European Community Growth No. GRD1-1999-10785, 2000-2003

[5] Dankwort C. Werner, Faisst Karl-G.: "Engineering in Reverse - A Holistic Extension of CAD", in: Papers of the workshop "New Trends in Engineering Design", Balatonfüred, Hungary, pp. 91-94, June 27-28, 2003.

[6] C. Werner Dankwort, "CAxSystems Architecture of the Future" in: Dieter Roller, Pere Brunet, eds.: "CAD Systems Development. Tools and Methods, , pp. 20-31, Springer 1997

[7] Ridwan Sartiono, Andrzej T. Janocha: "DESICA - Design in Context in der Aggregatekonstruktion bei Volkswagen AG", Conference "Informationsverarbeitung in der Produktentwicklung 2001: Effiziente 3D-Produktmodellierung - Fortschritte und Fallstricke", (CAT Engineering 2001), Stuttgart, 19. und 20. Juni 2001

10 Author

C. Werner Dankwort, University of Kaiserslautern, Dep. of Mechanical Engineering, POB 3049, D-67653 Kaiserslautern, Germany, e-mail: dankwort@mv.uni-kl.de

Panel Discussion

Summarized by Christine Hoffmann[1] and Andrzej T. Janocha[2]

1) Christine Hoffmann, Technische Universität Kaiserslautern
2) Andrzej T. Janocha, Freelance PLM and CAx consultant, Kaiserslautern

Participants:

> Umberto Cugini, Politecnico di Milano
> Rainer Stark, Ford Werke AG
> Werner Dankwort, Technische Universität Kaiserslautern
> Alfred Katzenbach, DaimlerChrysler Research and Technology
> Harald Robok, IBM-PLM-Solutions
> Akihiko Ohtaka, Nihon Unisys

The panel discussion started with participants' statements about the CAx and PLM future within the next decade.

Mr. Katzenbach recalled the history of CAD/CAM, CAE with its long struggle for sufficient data exchange. He mentioned all the efforts about data interfaces over the past twenty years. In his opinion, it would be a great progress, if there will be a final solution of this deficiency.

Mr. Stark talked about the term of interoperability in a broader context. He mentioned the interoperability between vendors, which is difficult to achieve, but – from his point of view – much better than four or five years ago. Regarding the interoperability between the different modules of UGS PLM Solutions, he demanded more harmony between the modules. Finally, he pointed out the interoperability between human beings and the systems as a very important issue, where, at the end, the difference in bringing business benefit will be done.

Mr. Robok's statement concerned two questions: that one of the business drivers in the PLM world and that one of the productivity of the end-users. He identified the human capital of the company as the most critical business driver, as the high skilled engineers deliver high value in the processes but also cost a lot compared to the worldwide team. It is a challenge to leverage this potential in the product development at its best. This can be achieved by two major issues: an evolution of the ergonomics of the User Interface (the human interface), which should be able to accept natural language, taking the engineer's language and transform it into the "language of the system" and, on the other hand, by better use of the possibilities of the systems.

Mr. Ohtaka pointed out the big economy loss caused by insufficient collaboration between end-users, IT vendors and researchers. He demanded, that in the future the end-users should better clarify their essential issues, with priority and conditions to solve them, researchers should better clarify what kind of theory or what technology is promising for solving this issues and IT vendors should propose ideas and specifications for better implementations of those issues. This very close collaboration between these three parties is essential.

Mr. Cugini stated the enormous interoperability problem, which is the main limitation in using IT technology today. He cared out about the way, knowledge should be modelled, formalized and handled in an independent (non-proprietary) way by end-users, vendors and researchers to avoid the interoperability problem in the future, even though this demand causes a kind of conflict and altercation today.

Finally Mr. Dankwort mentioned that the realisation of innovations usually take much more time than expected and even "short-term" means a period of about five or six years in this context. So on the one hand innovation cycle time should be reduced and on the other hand we have to think about future developments now. We should consider that our knowledge is only conserving things of the past, but what is really needed is some kind of "computer aided creativity". To set up these ideas, a new quality concerning the cooperation between engineering and economic should arise.

Several suggestions from the audience followed. Mr. Philippe Laufer (Dassault Systemès) suggested to "make innovations happen more quickly" by taking a couple of studies and trying to understand why they didn't come to the market quickly – if they weren't robust enough, the technology wasn't ready or the communication between researchers and vendors weren't established well enough. Maybe the separation in an isolated research lab that doesn't have to care about industrial pressure but is only studies new technologies and finally declares an innovation to be mature, and application developers to make the technology industrial would help. But, as Mr. Stark commented, there is no debate like "do we need technology?" because it just gives us new capabilities. The real question is how to introduce it, which means to think about things like "how do we want to penetrate the digital solution in our companies" and "whom do we prepare all this solutions for"?

Mr. Katzenbach suggested fluctuating people within or even between companies for the time of about two or three years to make new experiences and learn to see things from another perspectives. Mr. Ohtaka demanded a paradigm change from the technology driven approach to more real task oriented approach in that the end-users better clarify their needs.

Mrs. Jivka Ovtcharova (University of Karlsruhe) moaned about the lack of cooperation between vendors, researchers and end-users. Source of this problem could be the different understanding of "innovation", because the researchers just want new research areas, the vendors more functionality and the end-users more business. So the question is, how to bring these groups together. While this is usually done by European projects, Mrs. Ovtcharova thought that a consortium is too limited and too bureaucratic. So she suggested to do some contract-research as it is done successfully by American universities. The advantages are that you have independent and neutral partners like universities, which are at the latest state of the art and are really enthusiastic.

Mr. Christian Weber (Saarland University) asked the "meta question" concerning innovation driving: "who is having the ideas for whom?". Shall business people generate them for the technicians, shall the researchers do this for the IT people? Should the bottom-up or top-down approach be considered?

Mr. Katzenbach added that we shouldn't ask the customers for their needs directly, but "go to the customers, investigate them, build our own mind and then innovate."

Mr. Geist (ConWeb) suggested thinking about the new dimension that's coming up by the global outsourcing, because many engineering work is given to eastern European countries or e.g. China, but the CAD philosophy is still very centralistic. So the question is, how future CAx tools are going to handle a community of global engineering collaboration.

Mr. Pregnitzer (PTC) remarked, that more openness between vendors is possible and already exists. For example there is an agreement between PTC, Autodesk and

UGS that makes it possible to share code and use each other's hotline to solve interoperability problems. Mr. Stark concluded that vendors nowadays seem to be more open-minded to discuss different concepts.

Mr. Cremers (BMW) commented, that when we are talking about system innovation, as the rule we are thinking about functionality and systems. Another aspect, which is even more important, is the issue of the "process" of knowledge transfer. He identified at least two such processes to be important: First, how to provide vendors, IT people, with the understanding of the end-user's processes, and how flexible are the systems can handle these processes. Second, how can we integrate the different systems in the environment of our companies? Also the IT people have to understand their own integration process, for example what does specific interfaces do and how can we manage the different release processes of approx. 700 different applications in the development process of an automotive company. How can we handle each year one or two releases of each system? That is also are very big challenges and should be discussed in the next workshop.

The discussion closed with a couple of short statements and "wishes" issued by the participants of the panel. Mr. Cugini wished, that vendors and IT engineers refocused on the system and the processes. Mr. Stark followed by wishing that they saw more the opportunities than just the issues. While Mr. Ohtaka wanted to have a system that can support "real design and thinking", Mr. Robok put this wish in a concrete form by asking for a better front-end that provides the right business model and support mechanisms. Mr. Katzenbach thought that the focus should be put on the people who are using the system. Finally Mr. Dankwort wished for more interoperability at any level i.e. between the people ("minds"), the companies, vendors, engineers and end-users, and, last but not least, the systems.

Authors

Christine Hoffmann, Technische Universität Kaiserslautern.

Andrzej T. Janocha, Freelance PLM and CAx consultant, Kaiserslautern,
a.t.janocha@plm-consultants.com.